ASSOCIATION FOR RESEARCH COMMUNICATION HABITAT ENGINEERING AND ARTS
居 住 工 程 与 艺 术 研 究 交 流 协 会
LAURA ANDREINI  MARCO CASAMONTI  SILVIA FABI  GIOVANNI POLAZZI

# SUSTAINABLE
# LANDMARKS

## 可持续性地标建筑（上）

石大伟 主编

中国林业出版社

# CONTENTS
# 目录

# ARCHITECTURES / 建筑设计

INTERIORS / **室内设计**

SETS / **装置**

APPENDICES / **附录**

# FROM THE CUBE TO THE ARTICULATED BODY

## Twenty years of investigations by Studio Archea

## François Burkhardt

One of the most important exhibitions in the architecture world is the Venice Biennial of Architecture. For the sixth edition in 1996, the theme of which was "the architect as seismographer," the exhibition committee and the curator had decided to devote the Italian section to the generation of architects who were then between thirty and forty years old. Given the general theme of the exhibition, the choice proved especially judicious. The goal of this selection was to determine which artists could follow in the footsteps of the masters of the preceding generation. In retrospect, we can see that the selection revealed a panel of personalities and groups of creators who in time turned out to be first class. It was exceptional to meet an elite of young talents who would go on to be trend-setters in their field; even more rare, these artists were culturally engaged in the debates of contemporary Italian architecture, through participation in the media and university instruction. Among them, the Studio Archea of Florence, Alberto Cecchetto, Paolo Piva, Cino Zucchi, etc.

Studio Archea, a group founded five years earlier, had only completed a few projects; of these, it was the Stop Line recreation center in Bergamo – the group's first project, as it happened – that brought them national attention. Their first efforts were very eclectic, as evidenced by the three projects that the committee chose: the Stop Line recreation center in Bergamo, an immense container resulting from the transformation of a warehouse, covered in perforated plates of Corten steel and destined to become, after a reconstruction that emphasizes mobility, a center for performances, conferences, and meetings; the project of systematization of a public square in Merate, whose design as a series of bastions and terraces takes up the concept of an integration of the perimeter of an ancient city in the new architecture; and finally the project of a new factory in Massa.

### A Fun Palace

Stop Line (1993) is a work that deserves consideration. It remains to this day Archea's only completed work of its kind, but already reveals an impressive maturity, by its sober construction certainly, but also by its reinterpretation of Anglo-Saxon avant-garde projects in this area. Stop Line doesn't fail to resonate with fans of the English fun palaces of the fifties, those entertainment and recreation centers begun by Cedric Price that marked the first techno-pop projects of the Archigram group. Thirty years later, Stop Line still displays clear signs of this influence, particularly in the alignment of stairways leading to the gallery and placed against the façade. Marked by a culture of mobility, that solution induces the gaze to seek almost instinctively the "crane" elements of the English avant-garde of the same period. In this particular case, the references to art are diverse. If the pop montages were already informing the standard-setting projects, Archea also cites Lucio Fontana by utilizing perforations on the façade, as well as the accumulations of Arman on the inside of the building. These references come directly from the culture of one of the group's members, whose family runs an art gallery and therefore has an intimate familiarity with contemporary art. Stop Line therefore reveals itself as a strong and well-balanced mix of structuralism and art brut, with light and color effects guaranteeing a visible identity to its designated functions. In all the work of the Studio Archea to date, only the Tango discotheque (2007), built in Beijing, takes up a similar language in the façade. The façade-sign – or better yet, the signs integrated into façades – became a constant reference in their work as time went on (consider the letters engraved in the cement wall of the public library of Curno 1996, the semi-transparent aluminum wall-curtains of façades in the town hall of Merate 2005, or the Nembro library 2004 with its ceramic "books" attached to the façades and serving as sun-blockers).

# 从立方体到建筑实体
# 阿克雅建筑师事务所

二十年研究

**弗朗西斯·伯克哈特**

在世界范围内的建筑领域，威尼斯建筑双年展是最重要的展览之一。1996年举行的第六届双年展的主题是"作为地震专家的建筑师"，展览委员会和策展人决定将意大利展馆用于集中展示年龄在30到40岁之间的建筑师。除展览的一般主题之外，事实证明，这一决定具有特别的远见卓识。这种选择的目的是要确定什么样的艺术家可以继承前辈精英的足迹。回顾这些年，我们可以发现，这一选择挖掘了很多优秀的建筑师和创造者集体，时间也证明他们是一流的。年轻的天才精英同时也是这一领域内的潮流引导，这种情况是很少见的，更少见的是，这些艺术家们通过媒体或是大学的教学从文化的角度对意大利现代建筑学展开辩论。这些团体和个人精英就包括佛罗伦萨的阿克雅建筑师事务所、阿尔伯托·西凯托、帕奥罗·皮瓦、西诺·祖奇等。

阿克雅建筑师事务所在此次双年展的前5年成立。那时已经有过一些建筑设计经历，贝加莫"停止线"娱乐中心就是其设计的第一个项目，这个项目使他们受到了全国范围的关注。最初的做法是相当折衷的，展览委员会所选择的3个项目就可以证明：第一个项目是贝加莫"停止线"娱乐中心，该中心是由一个仓库改造而成的大型建筑物，用考登钢多孔钢板所覆盖，在通过一系列注重移动性的重建之后，将用于建成一个集表演、大型会议和小型会议于一体的中心；第二个项目是对梅拉泰一个公共广场进行功能分类，其设计尤如一系列堡垒，将老城边界通过新群建筑的屋顶连接起来；第三个项目是马萨的一个新工厂。

## 娱乐宫
"停止线"(1993)是一个值得关注的作品。截至到今日，它仍然是阿克雅完成的唯一作品类型，它通过其冷静的建筑风格以及对盎格鲁撒克逊建筑先锋派的重新诠释，表现出了令人深刻的成熟印象。"停止线"项目并未偏离50年代英国娱乐建筑的娱乐性，此类娱乐和休闲中心起始于Cedric Prince，它是Archigram的第一个技术大众化项目。30年后，"停止线"依然具有自己的影响力，尤其是它通往画廊并设计在正面的楼梯。这是一种流动性的设计，这种设计方式本能地追随了先前英国先锋派吊架式设计潮流。

在这个项目中，对艺术门类的借鉴是多样的。如果说流行的蒙太奇手法已经适用到这些标志意义的项目当中，那么，阿克雅同时也通过在建筑正面打孔来运用卢西奥·丰塔纳的风格，同时在建筑内部运用阿曼的叠加手法。上述手法的运用直接来源于事务所中一位成员的文化背景，因为其家庭经营画廊所以对现代艺术非常熟悉。因此，"停止线"项目反映了一种强烈的结构主义和原生态艺术的巧妙混合，通过光线和色彩反映其设计功能之间的差别。在阿克雅所有的作品当中，只有北京的糖果KTV (2007年) 在建筑方面采用了同样的建筑语言。建筑正面的标志（或者更准确地说：融合到正面的各种标志）已经随着时间的推移成为了它在作品中经常使用的手法（如1996年科诺公共图书馆墙面混凝土中雕刻的字母；2005年梅拉泰市政厅正面的半透明铝制幕墙；以及2004年纳姆罗图书馆正面用于遮挡阳光的陶"书"）。

## Toward a history liberated from its historicism

Like other constructions in progress, the Merate square and its underground garage (1997) are a project that evokes a major concern of Italian architecture, its rich culture and unique legacy of historic monuments; how to integrate a modern architecture remains one of the fundamental issues for the architecture of masters such as Scarpa, Gardella, Albini or BBPR. The city of Florence itself is one of the centers for this type of work, with Michelucci leading the way, followed by Savioli, Ricci and Dezzi Bardeschi, who place links to the past at the heart of their reflection to get to an architecture that integrates with the contemporary, each in his practices and according to his style. The Florentine avant-gardes of the late sixties kept a certain distance from this tendency, but the theme of integration remains central. The following generation, which includes Archea, saw the addition of new concerns: the urban landscape, and especially the peripheries and the quality of housing. These were the two aspects that characterized the collective's work, to which we'll return later. The solution chosen for the square in Merate was born of this idea of bastion-terraces that, itself, was inspired by the conjunction between the historic center and an original terrain on an incline implying a higher platform as a point of view on the landscape. This evocation of ancient murals in a place where there weren't any before extends the historical center beyond its walls by adding a new part to the preexisting parts of the old city. This is what the group called a "history liberated from its historicism."

## Organs and Bodies

If the works of the Studio Archea often distinguish themselves, especially in area of housing, by a study of volumes that generate articulated interstices and therefore form a homogenous and living ensemble – a sort of collection of organs connected to each other to represent a body – the group also made reference to the medieval city, asymmetrical and articulated, horizontally as well as vertically. By contrast, the objects conceived separately outside of urban centers are generally composed of geometrical volumes. Between these two approaches is the project of the re-qualification of the via Tirreno in Potenza. Fifteen years of investigations led Archea to a new conception of architecture. Their commissions focused on residential architecture. The projects had to be integrated into the preexisting combinations of buildings, like in Tavernuzze (2002) or in Cava dei Tirreni (2007), in industrial zones converted into residential zones – like the conversion of the old Fiat factories in Novoli (2002) – or placed just outside the city walls like in Como-Ticosa (2007). These projects all have in common a way of distributing dwellings on lots in small groups of buildings linked by internal streets or little squares, thereby creating internal pedestrian zones that cross the constructed areas horizontally. In this way, Archea organizes as much space as possible on the ground for the constructed volumes, which makes it possible to reserve the upper floors for living, while leaving at ground level just the space necessary for access to the apartments. The remaining volumes at ground level, thus liberated, are distributed proportionately among commercial spaces and public passageways. The plan for the ground floor of the Borgo Arnolfo in San Giovanni Val d'Arno exemplifies this type of architecture. In Novoli (consider the conversion plans for the FIAT factory) the buildings were placed in such a way as to create narrow intermediate spaces, corresponding to the necessary distance between the façades of each block, a distance reduced according to the designated reference of medieval architecture, but remaining very articulated to allow perspectives that ensure the vitality of the entirety of the architecture and the high quality of the intermediate spaces. Another direction Archea explored was the establishment of intermediate spaces for parks and gardens, like in Como-Ticosa. The organic aspect of these groupings doesn't aim to reveal itself through visual concepts that are close to nature, but through a structural analysis of relevant bodies, which remain natural in their grouping and give the spaces a human dimension.

## Organic Bodies between Foreign Bodies

One could think that architecture of this sort is opposed to the preexisting constructed bodies in the environment but, as the following projects reveal, it will instead carry on a dialogue with these buildings with which the new constructions must integrate, and even completely revive their context. The re-qualification of the via Tirreno in Potenza (in collaboration with Enric Miralles and Benedetta Tagliabue / 2001-2009) fills a need for the rejuvenation of a run down urban neighborhood by recovering the space between two endless bands of nine and eleven story buildings that, until the project, formed a monstrous gorge reserved for vehicle traffic. Studio Archea restructured the gorge by rerouting the traffic in such a way as to gain buried, half-buried, or above-ground spaces, that the architects then redistributed along the length of a new route created by opening a sort of landscaped park, but above all by establishing a green zone for rest and meeting strewn with "rock formations" of reinforced concrete. Commercial spaces and terraces were also added, the whole allowing a lucky integration of stairways leading to different levels. The general conception of this construction is a diversification of levels on the entire descending route, both along the length of the park and behind the buildings, by constructing a series of walls

**摆脱历史决定论 创造历史**

与这段历史时期的其他建筑一样，梅拉泰广场及其地下车库（1997年）引起了人们对意大利式建筑的强烈关注，其丰富的文化、独特的历史遗迹；对于诸如斯卡帕、加迪拉、阿尔比尼或BBPR建筑事务所等建筑大师而言，如何将现代建筑融入其中，仍然是其建筑设计中的一个重要问题。佛罗伦萨城的建筑就是以这项工作为中心的，该城市的建筑设计由米凯卢奇牵头，随后有萨维奥利、利奇和戴奇·巴德奇加入其中，他们通过建筑设计来实践自己的风格，着重将老式建筑与现代建筑融为一体。60年代晚期佛罗伦萨的前卫建筑与这种潮流保持了一定的距离，但融合的进程仍然是主流。

包括阿克雅在内的新一代建筑师认识到了新元素的作用：城市景观，尤其是建筑的边际线和品质感。上述两个元素是事务所作品的标志，我们将在下文进行讨论。依据斜坡式地形而为梅拉泰广场所选择的设计方案，其灵感来源于这个历史性的中心和原有的地形，原有的地形在整个视野内形成了一个较高的平台。这种在历史中心再建新建筑的方式，通过向旧城原有的部分加入新的元素，使历史中心得到了扩展。事务所团队将这种做法称之为"摆脱历史、创造历史"。

**机体与整体**

如果说阿克雅建筑师事务所（尤其在建筑设计领域）通过制造衔接缝隙来显示自己的特色并因此形成一个同质的和鲜活的整体的话（建筑中的各个部分彼此相连，最终形成一个机体），那么，事务所团队也从不对称、衔接、水平和垂直等角度参照了中世纪城市的建筑风格。相反，在城市中心之外的单体建筑则通常由几何图形组成。介于这两种做法之间的建筑作品是波特扎蒂雷诺的重建项目。

阿克雅建筑师事务所15年来的研究，使其形成了一种新的建筑观念。它的项目着重于民用建筑。其项目必须与原有建筑融为一体，就象塔沃努兹（2002年）或卡瓦·德·泰瑞尼（2007年）那样，将工业区转换成居住区（尤如在诺沃利菲亚特老工厂的转变），或者如科莫提科萨（2007年）直接将其安排在城市围墙之外。

这些项目有一个共同的特点，就是将居住区分散于少量的与内部街道和小广场相连的建筑之中，以此在建筑区域内的水平方向形成一个跨区域的步行空间。依此，阿克雅在地面上尽可能地给建筑物留出了空间，使得将上层空间用于居住，底层留出的空间则可以使人们进入建筑之内，而其它部分就被解放出来，可以分区域用于商业和公共通道。

圣·吉尔瓦尼·瓦尔·德阿诺的博戈·阿诺尔福的底层设计就充分发挥了这种建筑风格。在诺沃利（参见菲亚特工厂的转换项目）建筑物的建设是用来实现一个狭窄的过渡空间，它的位置取决于各幢建筑立面之间的距离，但是却保留了衔接的视点，展现了建筑整体的重要性以及过渡空间的高品质。阿克雅探索的另一方面是为公园和花园建造过渡空间，如在科莫提科萨，这些组合方式的功能并不是为了展示其亲近自然的视觉概念，而是通过对相关建筑实体的结构分析来展示自己，而这些建筑实体依然保持着其组合的自然属性并且提供了更人性化的空间。

**新建筑之中的有机建筑**

有人可能认为，这种类型的建筑会与环境中原有建筑相冲突，但正如下述项目所示，新建筑反而会与原有建筑相呼应，新建筑与原有建筑融为一体，达到完全和谐。

位于波特扎的蒂雷诺路的重建（与Enric Miralles,Benedetta Tagliabue合作/2001-2009）满足了城市区域再造的要求，使其恢复了年轻的光彩。对于超长的九层和十一层沿线之间的空间，它进行了重新设计。此前，这一空间是造成车辆拥堵的巨大源头。阿克雅通过重新设计交通路线对这一区域进行了重新规划，开发地下、半地下及地面空间，通过一个景观公园重新分配路线，最重要的是，通过强化混凝土制成的"石头阵"建立了一个用于休息和聚会的绿色空间。同时，它还添加了商业空间带，通过整体将通往不同楼层的楼梯结合在一起。这种建筑方式的概念一般是通过为机动车预留的新通道旁边建造很多墙，扩展在公园边和建筑物之中不同楼层及下楼路线的多样性。除了恢复原有街道的公共空间，这种设计方式的重点在于视觉的构建，使得人们在通过这一区域时可以看到一系列连续的开放或封闭的空间。

卡瓦·德·泰瑞尼老旧烟草工厂（2007年）的设计就是为了通过重建老工厂和加入商业、居住元素来使这块土地获得公共空间的功能。在此项目中，建筑师提出建造一个长的Z形建筑，地面空间的一部分可以自由通行或用于商业目的。但是，这座大厦的空间密度无法实施阿克雅所提出的设计方案，如用不同建筑配比来保证更多的自由空间组合并建立一个过渡地带。这种想法在本案中未能实施。

在米兰的托瑞·德里·阿提项目（2008年）中，阿克雅事务所团队试图将新的标志融入城市风景之中，为米兰城建造第三个

along the new path reserved for motor vehicles. Beyond the recovery of a public place from what had been a street, the interest of this operation resides in its visual structure, which forms a segmentation of the space with respect to its lateral buildings and creates a continual succession of different open and closed places that one discovers in turn while passing through the area. The old tobacco factory in Cava dei Tirreni (2007) was redesigned so that the lot regains its public space functionality by reconverting the old factory and by adding commercial and residential functions. In this case, the architects proposed a long Z-shaped building, with part of the ground level remaining free, or occupied by the commercial spaces. The density of this long edifice, however, does not allow for the solutions usually proposed by Archea, like the pairing of different buildings to guarantee a greater freedom of composition and create intermediary spaces, which are missing in this case. With the Torre delle Arti project in Milan (2008) the collective attempted to integrate a new sign into the urban landscape, to offer the city of Milan its third high-quality vertical architectural work. Complementing the Pirelli tower by Gio Ponti, still in the tradition of modern architecture, and the Verzasca tower by the BBPR, which is aligned with the neo-historicism movement of the late fifties, Archea's Milan Torre delle Arti aims to express in its vertical structure the plastic qualities that the architects managed to integrate so successfully into the functions of the horizontal buildings. In this case, the problem is slightly different; the fluid articulation of the elements, so clear in their search for horizontal coordination, cannot be expressed so easily through verticality. In fact, the strong articulation given to the volumes and the horizontal lines, in addition to the different height of each part of the Torre delle Arti, elicits a hyper-articulated expression, an effect which is further reinforced by the various types of openings on the façade. This tends to compromise somewhat the fluidity of the relationship between the body and the organs of the building. The solution that was considered for this problem – a conjunction of the arts and languages derived from electronic technologies and transposed on the façades – underlines the expected effect of integration with the architecture itself and cancels out the organicism that would have made it too artificial in a vertical edifice, contrary to the success the architects had in their organic investigations for horizontal constructions. This endeavor seems interesting and courageous to me, and deserves to be pursued. The last project selected to conclude this chapter on the coexistence of organic bodies and foreign bodies is the Beijing Parkour (2007), a project conceived for an exhibition presented in China. With this project, Archea confirmed its talent for integrating a new architecture within a preexisting urban grid: the concept was based on the transformation of the traditional hutong habitat, while preserving the perceptive memory. Again, Archea focuses on creating multipurpose routes, which encourage exchange and social interaction. Using a kind of underground network that connects all the residential cells and includes basic public services, the artists return the public spaces to their inhabitants, by eliminating the barriers created by traditional separations. The architecture of the project itself is a benchmark for the renewal of local tradition. History remains an important aspect for the Studio Archea, inasmuch as the project submitted to them provides the valences necessary.for renewal.

## Some Solitary Objects or the Play of Surprises

The most recently completed work is the Residenza del Forte Carlo Felice in La Maddalena, near Sassari (2009). It's an immense hotel complex situated on a seashore, with a natural zone as the backdrop and the historical monument of the Residenza Carlo Felice, also facing the sea, at its center. Studio Archea's project focuses on two dominant aspects of the landscape: the historic residence and the relationship with the natural site. The constructions seek to integrate as discreetly as possible, while at the same time distinguishing themselves as the expression of a strong and identifiable architecture. Here again it was Archea's facility with the articulation of volumes – kept very low with respect to the original historic monument that they so skillfully surround, – that structured the complex by recovering the spaces between the harmonious and intimate blocks. The different bodies of dwellings are marked by deep verandas, and the walls of apparent stone typical of the region ensure a good integration, as do the flat rooftops covered in grass. The basic idea is to make the architecture, seen from the perspective of the hill in the rear, disappear in favor of the natural landscape near the sea, by integrating the restaurant placed behind the complex and giving it an underground entrance. This contemporary architecture rich in elements of integration invents an artificial landscape perfectly integrated with the natural landscape, an example that could certainly be instructive in Sardinia. For the competition of the Museo d'Arte Nuragica e Contemporanea in Cagliari, in association with Franz Prati e MDU Architetti, Archea presented an extremely interesting project. The authors successfully united two fundamental aspects of the functions that a building in this location should possess: in conceiving the museum, they had to create a new distinguishing sign on the length of the coast while making its role apparent, especially by means of volume, but also to recall the local prehistoric culture of the nuraghi. To do this, they had to remodel the landscape where the museum was to be built, a veritable urban project of landscape design that is at the same time indicative of the local identity. The nuraghi, monolithic constructions, include an internal room with a corridor that leads to the outside by a door with an architrave. The Archea project reinterprets this original conception. Five floors connected to

优秀的竖向建筑作品。此前，吉奥•庞帝设计的倍瑞力塔仍然属于现代建筑，BBPR工作室设计的韦尔扎斯卡塔反映了50年代末期的新历史主义，而阿克雅的米兰艺术塔项目，则是要表达竖向建筑的硬质感，将其与水平建筑的功能成功地融为一体。在这一项目中，所出现的问题稍有不同，其在水平结合的探索中所轻易得出的各种元素的流动性，却在竖向建筑中无法轻易表达。事实上，除艺术塔建筑各个部分高度不同外，建筑组成和水平线条的强烈形象，形成了一种超形象的表达，这种表达更通过建筑正面各种类型的开窗得到了进一步的加强。这种做法在某种程度上妥协了建筑的实体与部分之间的关系。对这一问题所提出的解决方案重点在于建筑本身一体化的效果预期，抵销了在竖向建筑中可能添加过多人工成份的有机论，这与建筑师们在针对组织功能的研究中所取得的成功完全不同。总的来说，这种努力似乎很有趣，值得去尝试。

本章中，最后一个有关有机建筑和新建筑所选择的项目是北京跑酷（2007年），该项目是为中国的一次展览而设计。在这一项目中，阿克雅再次证明了自己的天赋，它将新建筑与原有街区相结合：这一概念的基础在于将原有的胡同空间进行转型，同时保留空间记忆。阿克雅再次将主要精力集中在多功能的路径设计上，确保交互性和社会互动性。通过一种地下网络将所有居住区域连接在一起并将公共服务职能结合在一起，建筑师通过打破原有的区域界限，将公共空间回归给了居民。这个项目成为了当地传统再造的基准。对于阿克雅事务所来说，历史是非常重要的，接手该项目就已经开始在重建中关注历史问题。


## 一些特别建筑或追求新异的表演

最近完成的一个项目，萨萨里附近位于拉马达里纳的卡罗•菲利斯堡酒店（2009年）。这是一幢大型的海岸综合酒店，同样面朝大海的卡罗菲利斯堡历史建筑位于自然背景的中心位置。阿克雅事务所的建筑设计主要集中在环境中的两个重要方面：历史建筑以及其与自然景观的关系。

阿雅克设计的每个建筑试图尽量谨慎地与周边环境融为一体，同时通过其强烈而独特的建筑特色将自身与周边区别开来。这次阿克雅将酒店巧妙地围绕在历史建筑的旁边，对其造成非常小的影响，在和谐而又紧凑的街区之间恢复了空间，并对复杂的建筑进行了重构。住宅的不同实体之间有着标志性的走廊。由当地特色石块所建造的墙壁确保了良好的融合性，掩映在芳草中的水平屋顶同样融于周边环境。其基本想法是通过在建筑组合后面连接餐馆并且设计一个地下通道，来使该建筑（以背后的小山为视角）消失在海岸的自然风光之中。这个现代建筑将很多的人工景观与自然景观完美地接合在一起，当然这种想法也对撒丁岛的建筑有所借鉴。

在卡利亚里努拉吉艺术与当代艺术的博物馆竞赛中，阿克雅与Franz Prati e MDU Architetti一道提交了一项非常有趣的方案。设计师成功地将本地建筑功能中的两个重要因素融合在一起：在设计博物馆过程中，他们必须在海岸线上设计一个新的醒目标志，明确地表达其功能，尤其是通过形状，同时还要融合当地努拉吉建筑的历史文化。为了做到这一点，他们必须重新规划博物馆建设地点的周围的景观，设计一个真正的城市建筑，同时能够反映当地的特点。

传统的努拉吉塔状建筑的内部只有一个房间，外有走廊，通过架有楣梁的门连接到外部。阿克雅的这个项目重新诠释了这一原始的概念。一个连续的，内置庭院的环形坡道竖向连接5层楼面，在纵向上拉伸空间，并且随着空间的上升逐步缩小，就像一个锥形的塔。水平的层次感唤起了当地建筑中特有的不规则石质阶层建筑风格。精巧的内部设计显示了阿克雅在处理整体规划、内部空间和外形之间关系的成熟度，同时也显示了其将建筑与环境高度融合的能力。

为了继续对建筑与自然景观之间的融合进行深入研究，阿克雅设计了San Casciano Val di Pesa的Cantina Antinori。这是一个反映这种融合的重要项目。建筑整体完全隐蔽于地下，只能在葡萄园处看见两个开口，其中一个为建筑的入口，另一个为垂直的光井为办公室提供光源。对于葡萄酒生产和储藏的空间，设计师设计了一系列节奏起伏的弧顶，该弧顶用砖砌成，裸露在外，与墙壁合为一体，连续而又紧密，展示了一种新的设计概念。需要再次说明的是，这个建筑作品的特异之处在于其出人意料的内部设计，由于外部无法看出建筑内部的复杂性，所以对于参观者来说是出人意料的。

对于阿雅克事务所的作品，我们所列举的案例会使读者了解其最初的概念原型。虽然只是部分作品，但体现出的设计理念还是颇具多样性的，并且能够看到一条清晰的主题，我们可以总结为以下几点：

1. 通过合理安排创造流动空间来激活建筑空间的力量。
2. 不断努力，尽量节省出地面的空间，以此来扩大公共空间和半公共空间，将它们与私人空间相连接，促成两类空间的对话。
3. 通过使用雕塑来激活空间。
4. 通过使用原有材料，给建筑正面加入新的组织元素。
5. 将历史作为重要的遗产，通过新的作品来复活现有的建筑布局。

one another by a ramp are grouped around a vast courtyard that spans the space vertically. The desired effect is that of very elevated central space, open to the sky, and marked by a sort of gradation that shrinks as it gets higher, an allusion to the conical cupola of the nuraghi. The horizontal stratifications recall the irregular superimposed masses of blocks of stone typical of that style of constructions. The mastery of the interior space shows a maturity that the Studio Archea acquired in the relationships between program, content and container, between spatial envelope and architectural body, as well as integration in the environment. Continuing their investigations into the integration between architecture and natural site, Studio Archea planned the Cantina Antinori in San Casciano Val di Pesa, currently under construction (2005-2010). This is a radical project in terms of that integration. The building will be entirely underground, incorporating a hypogeal and cryptic architecture that completely disappears inside the ground under which it is built, showing only two open incisions in a landscape covered with vineyards, one of the openings offering access, the other letting light into the offices, the other natural light sources being vertical light wells. For the part reserved for the production and storage of wine, the organism is characterized by a rhythmic sequence of vaults made from natural brick left exposed and united with the walls to create a succession of spaces that display both great intensity and a new concept. Once again, the particularity of this work owes to the unexpected modulation of internal spaces, unexpected from the perspective of visitors since there is no external sign of the complexity of the interior. This brief panorama of a selection of works by Studio Archea gives us a few telling indications of their original conception of an architecture that, despite its heterogeneity, nonetheless permits us to identify a clear theme composed of diverse aspects from which we can derive several criteria of specificity, even if these criteria are only relevant for part of the discussed works:

1. the ability to bring the architectural spaces to life through a skillful deployment of volumes that are coordinated in such a way as to create fluid spaces
2. a constant attempt to free up as much space as possible on the ground, in order to gain public or semi-public expanses and to connect them to the private ones, bringing the two into conversation
3. the use of sculptural-architectonic composition to revive the spaces
4. the use of preexisting materials to try out new textural elements in façades
5. the consideration of history as a valuable legacy, while the new work contributes to a liberation from the status quo
6. the importance of integration in the surrounding context, be it an urban setting or a landscape
7. the creation of surprise effects by means of a confrontation between internal and external volumes that do not necessarily correspond to one another
8. the active participation in the creation of an organic architecture movement, iconographically liberated from references to the vegetal world and to animal and human skeletons
9. an articulation based on a rotation of plans, cross-sections, and façades, by integrating diagonals that constitute new interfaces.

These criteria underscore the originality of twenty years of investigations and give Studio Archea a specific position in the panorama of new Italian architecture.

6. 强调建筑与环境景观结合在一起的重要性，无论其为城市景观还是自然景观。

7. 在并不必然相关联的内部和外部形状之间制造冲突，创造一种意想不到的效果。

8. 积极参与功能建筑运动的发起，脱离效仿植物、动物和人类骨骼形象。

9. 通过斜角制造新的连接界面，并通过设计图、变换部分和建筑正面的变化来完成转换。

上述特色记录了阿克雅建筑事务所20年来的原创研究成果，确立了其在意大利新式建筑领域的特殊地位。

# ARCHEA: NON-STYLE DESIGN AND INTERPRETATION

## Fang Zhenning

I came into contact with Archea Studio in 2006, shortly after the opening of the Beijing branch. What impressed me most on that occasion was a project then in the planning stage, for the Antinori winery in Casciano Val di Pesa, which enabled me to appreciate the design acumen and high quality standard of Archea. It is no coincidence that when we met four years later, our discussion commenced with this project.

### Observation and dialogue

In the summer of 2010 I visited several projects developed by Archea in Milan and Florence, concluding my research journey with an interview with Marco Casamonti at the headquarters of Archea.

In the last ten years I have had the opportunity to meet some of the world's leading architects and artists, but it was the first time I interviewed an Italian architect. Italy is the country where I have travelled most, I am always curious about what contemporary Italian architects think. In a period dominated by globalization, regional culture is assuming an unprecedented importance, and the projects of Archea are very illuminating examples of this. I began to truly understand the design philosophy of Archea in early September 2010, when I arrived in Milan. I have always paid great attention to the relations between materials and colors, on the one side, and the regional and historical context on the other, in their projects. While I have certainly been very impressed by the monumental quality of their architectures, after having seen them and discussed them with their architect, I have begun to fully appreciate the work of Archea.

Contextuality and strategy

Marco Casamonti and his partners Laura Andreini, Giovanni Polazzi and Silvia Fabi take inspiration for their projects from history and from a profound love for art. When observing the buildings designed by the Studio, one clearly perceives the heritage of the aesthetic principles that were already applied by the ancient Romans. Spatial conformation and measure interact with results that can only be defined as magnificent, as in the case of the former Manifattura Tabacchi□2007), a landmark from former times that has become a new and innovative townscape. I have asked myself whether Italy has more ancient buildings than other countries, or is there a strong tradition to preserve and maintain traces from history? Well, I believe the latter to be the case.

I also wondered why the projects of Archea comprise so many renovations. Marco Casamonti has dispelled my doubts, explaining that the firm prioritizes works which do not occupy new land, because the territorial resources and the energy they represent cannot be recreated. The construction of new buildings cannot but compromise the territorial heritage. The architects of Archea are therefore determined to implement a strategy aimed at preservation, which does not damage the environment and respects existing architectures in order to preserve the land which is still free, in the best possible manner. Marco Casamonti has adopted this philosophy, in the conviction that territorial resources are not renewable, and that the use of land for new buildings should therefore be reduced.

There is no shortage of land for building sites in Italy, but the commitment to reduce the use of land is a sign of the awareness of an architect, witnessed through ethical choices.

### Non-style and interpretation

Both in history and in contemporary architecture one tends to judge a work of architecture on the basis of its style. The projects of Archea, on the contrary, reject any concept of prevailing style. Marco Casamonti declares that he is not interested in the formal language and the assertion of a pure style through

# 阿克雅：
# 非风格设计与演绎

**方振宁**

2006年，也就是在阿克雅(Archea)建筑师事务所在北京落户不久，我来过这家事务所，当时给我印象最深的是一个正在建设中的项目，即位于佛罗伦萨San Casciano Val di Pesa地区的Cantina Antinori酒窖设计方案，这使我初次感觉到阿克雅的实力和水准，正是由于这个原因，当4年后我们有机会再次见面时，话题是从这个平台起步的。

### 考察与对话
2010年夏，我先后在米兰和佛罗伦萨参观了阿克雅设计的几个项目，最后以在佛罗伦萨阿克雅建筑事务所本部和马可 卡萨蒙蒂(Marco casamonti)对话的方式，结束了对阿克雅建筑项目的考察。
在过去的10年里，我有机会和一些世界建筑师和艺术家对话，可是和意大利建筑师还是第一次。意大利是我去过最多的国家，我对当代意大利建筑师思考什么一直抱有好奇心。尤其是在全球化的时代，地域文化的价值正在显示出前所未有的价值，阿克雅的设计在这方面给我们很多启示。
我对阿克雅设计真正的了解，是2010年9月初从米兰开始，我一直关心他们的项目所使用的材料、色彩和地域以及历史的关联，当然我也对这些建筑的纪念碑性印象深刻，可是当我考察了这些建筑并与主持建筑师深入交谈之后，才开始有了新的理解。

### 文脉与策略
马可先生和他的合作伙伴Laura、Giovanni和SilviaFabi，对历史的了解和对艺术的挚爱，是他们设计的出发点。从这些项目中可以看到罗马时期就确立的美学原则，透过尺度所体现出来的格局，它们显得相当震撼，比如为萨勒莫设计的Manifattura Tabacchi (2007) 就像是一座古代的遗迹，但却成为这座城市中新的风景和地标。是意大利的历史建筑比别的国家更多，还是意大利自古就有保存和维护历史建筑的传统？我认为是后者。
我对阿克雅的作品中很多项目是改造项目感到好奇，马可的回答解消了我的疑问。他说："为什么我们的设计项目中很多是老建筑改造项目？其原因在于，我们不希望建筑占用更多的土地资源。我们认为这种土地资源及其代表的能量是不可再造的。如果建设新的项目，就势必会占用掉一些土地资源。我们希望在不消耗土地资源和尊重老建筑的前提下，尽量使用改造策略，这是一种以最大可能性节约土地的策略。"他把这种观点当作一种主张提出来，理由是：尽量减少使用土地用于建设项目，这是由于土地的不可再生性决定的。
意大利并不是一个缺少土地的国家，但尽量减少使用土地成为建筑师的一种自觉的准则。这种伦理观对所有的建筑师来说都是启示。

### 非风格与演绎
在建筑的历史上和当下的建筑界，以风格来衡量建筑是常态，而阿克雅的设计却是非风格设计，马可对建筑的形式语言和设计风格并不感兴趣，他把风格设计视为"惯性思维"，这是一个很好的比喻，这也就使得阿克雅的每一个建筑项目都不雷同，出人意料。那么阿克雅的作品是如何演绎的呢？如果假设把某一个项目特点作为风格的参照系，那么它和其它的项目几乎没有任何联系和可比性，我们在阿克雅的设计中找不到和NEMBRO LIBRARY（书墙图书馆）有着血缘关系的设计，因

architecture, an approach he aptly defines as an "inert thought". This is why the architectures designed by the firm are very variegated, and often surprising. What is the key to interpreting the work of Archea? If we were to use a single characteristic of a specific project as stylistic criterion, it would become very hard to retrace links and resemblances with other works by the group. There is nothing comparable to the Library of Nembro in the entire portfolio of Archea. In this sense, interpretation becomes a chimera.

When I visited the old town hall of Merate (2001), renovated by the firm, I have recognized echoes of Arte Povera, a movement which developed in Italy in the Sixties, and when seeing the corridors and footbridges in steel sheet my intuition has been confirmed. Casamonti himself has declared that Arte Povera has been a source of inspiration for the group, imperceptibly permeating the planning works. The artists of the movement did not exalt poverty as such; it represented a means to rebel against the exclusive use of traditional materials in the production of works. Their art was made with everyday materials. Archea rarely uses expensive materials in its projects, preferring inexpensive ones as steel, wood and concrete, true to the basic ideas of Arte Povera. When the ideas of architects meet those of artists, the result is a spontaneous creation of new architectural forms.

## A new world and a new mission

Archea has several branches in Italy, and the opening of a Chinese office has undoubtedly been a strategic move. Marco Casamonti refers to the concept of nomadism to describe the identity of an architect. Nomads are people who are always on the move, who travel from one place to the next as the environment changes, and Casamonti considers China as today's country of change. We will now examine how the Beijing branch of Archea is developing its activities in the Far East.

The Archea firm has understood that the profound changes taking place in China may result in design opportunities. Italian architects, who can also count on an ancient heritage and who appreciate the importance of preserving architectural testimonials of their history, have launched an appeal to China. In fact, Casamonti considers that there is reason to be alarmed, because many Chinese fail to appreciate the culture, traditions, history and the distinctive traits of their country. The conduct of numerous real estate developers has resulted in the destruction of large areas. The principles of respect and protection of the local culture, which are deeply rooted in Italy, are particular significant in the case of China. The mission of Archea is, precisely, to diffuse a similar mentality in China.

Casamonti has been very explicit in this regard, but is this truly so? And if so, how does Archea think it may implement its philosophy and practice in China? One example for all is the renovation plan developed by Archea for a number of historical building complexes in Beijing. The "Beijing Parkour" project (2007) and Archea has succeeded perfectly in translating the concept in reality, basing the work on the self-same definition of "Parkour", the art of identifying physical and human spaces associated with an artistic value through rapid moves. Quadrangular court houses are a typical morphologic structure of the residential urban tissue of old Beijing. With the passing of time, the physical space of the court house has undergone changes and the debate on its architectural recovery has always been very intense. Numerous Chinese architects have participated in the debate, and have been joined by foreign architects in the last decade; each has suggested a different solution to the problem. The idea presented by Archea casts new light on the issue. Its strategic vision is based on a sustainable and integrated development of this historical building complex. One may object that anyone participating in renovation projects agrees with this approach. So what distinguishes the project of Archea from the others?

Archea has not recovered the quadrangular court structure by creating a fake antique, as many have done in an attempt to satisfy the psychological need for novelty while remaining attached to a futile nostalgia. Rather, the Italian architects have suggested an underground surface which, connected to the existing structure, frees up the labyrinthine tissue of the alleys in the third dimension, thus removing the barriers imposed by the traditional architecture and the enclosures of the court houses, at the same time preserving the proportions and the morphological structure.

Archea has also dealt successfully with another challenge associated with this kind of project, namely the preservation of spatial continuity, overcoming the distinction between public and private space. Many Chinese architects have failed to go beyond the mature style of the court house because this perfect configuration has become a psychological cage to them. Archea has succeeded in deconstructing the original fixed scheme, opening it to various possibilities of use and combining dwellings, shops and cultural venues in various elements that integrate the existing urban tissue. The project is organized in a three-dimensional complex of interconnected floors, connected by escalators and equipped with service facilities.

The historical motivations of the court house are familiar. But it has been developed in a period in which there were no motor vehicles, and thus the width of the streets and the dimensions of the public spaces no longer meet the requirements of communication and traffic. The optimization of vehicle circulation has therefore been a priority. The renovation plan developed by Archea has first and foremost entailed the deconstruction of the historical structure of the court house, in order to create streets accessible for vehicles and pedestrians which cut through the urban tissue on various levels, realizing a volume of 20,000 cubic meters on an area of 18,000 square meters. The three-dimensionality of the spaces has made it possible to maintain the original height of the building, preserving the profile and the historical memory of the city, without an excessive development in height.

此"演绎"就成为一个谜。

起初，我是在考察对旧市政厅进行改造的项目 – MERATE市政厅(2001)时发现里面有用钢板制作的走廊或者说是天桥的时候，意识到它和20世纪60年代在意大利兴起的有名的艺术流派贫穷艺术(Arte Povera)有关，最终，我的直觉得到证实。马可强调贫穷艺术确实对他们有很大的影响，艺术在潜移默化着他们的设计。创作贫穷艺术的艺术家并不贫穷，这只是他们的观念，只是对使用传统材料创作作品的一种反对，他们使用在日常生活中可任意取得的材料进行艺术创作。阿克雅在设计中很少使用奢侈的材料，而更多地使用廉价的材料，包括钢材，木材，混凝土，这正是贫穷艺术的主张。当建筑师在观念上和艺术家合流时，产生新的建筑形态是自然的事。

## 新大陆与使命

阿克雅在意大利有好几个分部，而在中国的阿克雅则是阿克雅战略性的一步。马可用游牧人(Nomadism)来比喻建筑师的身份，所谓游牧人就是总在向那些发生着巨变的地方迁移，他认为现在的中国就是这样的一个地方。那么让我们来看看在北京的阿克雅是怎样在远东开始展开他们新的事业。

作为建筑师的敏感，阿克雅意识到发生着巨变的中国，会给建筑师带来机遇，但来自另一个文明古国，和一向有着用建筑的形式尊重历史的意大利的建筑师，却对中国发出忠告，马可认为：现在的中国处在一个非常危急的时刻，就是现在很多的中国人都不尊重自身文化传统，以及历史和自身特点。很多地产开发商的行为对环境造成了极大的破坏。意大利对自身文化尊重及保护方法对中国很有现实意义。这也是我们来到中国的一个使命。

马可说得非常直白，那么是不是这样？如果是，那么阿克雅是以怎样的姿态来参与和把他们的经验用于中国？让我们看看阿克雅为北京的历史街区提供的改造方案，主题为"北京跑酷(Beijing Parkour)"的项目(2007)，实际上是假题真做。"跑酷"的意思是在以快速度浏览中发现和创造有艺术含量的空间或人文。四合院，是由于北京人以往的定居式的家族结构决定的居住形态，随着时代的变化和时间的推移，四合院的物理空间在分化，而关于它的改造问题，一直是一个议论的焦点。很多中国建筑师都参与了这一议论，而最近10年里，也有外国建筑师参与到这些议论中来，大家都提出了无数的解决方案。阿克雅建筑设计的提案显然是一个新的角度，他的战略性思考，是如何让这个历史性的街区能够深入和持续的发展下去。也许有人会说，凡是参与这一项目改造设计的人都会以此为出发点，那么阿克雅的高人一筹之处在哪里呢？

阿克雅不是像许多以往对四合院的改造那样，把四合院修复成一个假古董的状态，来满足人们对这一建筑形态的猎奇心理和廉价的乡愁。阿克雅的手法是：通过一个连接到现有构筑物的地表下区域，将胡同从迷宫似的网络中解放出来，让它进入到一个三维的世界，这样可以消除传统建筑类型中的空间障碍和四合院的蔓墙，同时保留它原有的比例和形态结构。我们从这个方案中看到了这些承诺。

这个方案还有一个根本性的挑战，那就是在空间上的连续性设计，它克服了私有空间与公用空间之间的严格界限。而在这一点上，许多中国建筑师的设计之所以难以逾越成熟的四合院样式，是因为这一完美的形态已经在中国人的心理上构成屏障。而阿克雅的做法，是解构了原有的封闭形态，从中开发出各种应用的可能性，将住宅、商用、文化和各种预制元素与现有建筑基本结构结合，由多个自由通行的楼层组成，配备了自动扶梯和设置了服务区，从而形成了一个三维尺度的建筑物。

其实我们都知道，形成四合院的历史因素，首先，那不是一个以车代步的时代，所以街道的宽度，也就是公共空间的尺度是不适合现代人的活动和交往的。因此优化车辆的通行能力，是对四合院改造下手的第一刀。阿克雅的改造方案，首先解构了四合院的基本结构，通过让车辆和行人通道在各个层面进行渗透的手法，将一个建筑面积只有18,000平方米的空间，优化为有着200,000立方米的体量。空间的立体化，使得我们能够不冲破原有建筑高度，这是为了保持城市的肌理和记忆，而不是为了形式上的等高。

# WORKS

设计作品

# PROJECT REPORTS

## ACOUSTIC BARRIERS
### Impruneta, Italy, 2000-2007

This architectural and landscape project is part of a comprehensive improvement plan for thc highway infrastructurc for thc third lane of the Florence-Bologna section of the A1 highway.
The project's primary purpose is to plan and build noise protection elements in addition to solutions targeted at environmental and infrastructural mitigation, such as transfer hub parking lots and artificial tunnels. The noise barriers, which are both absorbent, made of terracotta and Corten steel, and reflecting, in Corten steel and glass, were designed as architectural constructions that give character to the landscape along the highway section by their references to a traditional local material. The barriers are made of a system of microperforated Corten steel panels, supported by "double T" beams on which there are repeated linear sequences of terracotta elements. Beyond the effective sound reduction from the panels with rock wool inside, the terracotta from Impruneta withstands smog and frost, which greatly reduces maintenance costs and gives it the appearance of historic city walls.
**pag 76-83**

**location** Impruneta, Florence, Italy
**project** Sound-absorbing and reflecting barriers
**client** Autostrade per l'Italia S.p.A.
**structures** CIR-Ambiente
**plan** 2000
**construction** 2005-2007
**cost** € 6,500,000
**built area** 8,3 sq.km
**surface area clad in terracotta** 20,000 sq.m
**contractor** CIR-ambiente

## VIA TIRRENO REDEVELOPMENT
### Potenza, Italy, 2000-2010

The project area covers a volume bound by large apartment buildings in the working-class ncighborhood of Cocuzzo in Potenza. The space is that left between two rows of buildings, covered by a street that greatly limits the amount of free space on the ground, minimizing the availability of parking and nullifying the existence of greenery and pedestrian paths. Rejecting a mere cosmetic face-lift, the design takes the context as a conflict between the artificial environment, the "cement canyon" of the façades of the buildings across each other, and the natural environment, in which the space between the two windowed walls is treated as a ravine, a valley floor. This space is defined by boulders and rocks that give life to a large straight stretch of park that is free of cars and streets. The natural slope lets a new ground surface be created that brings some parts underground, hiding the parking areas and creating raised scenic terraces. On the ground, asphalt was replaced by a spreading green field, animated by a series of blocks that emerge along the promenade, multi-faceted masses of cement pigmented with iron oxides.
**pag 84-95**

**location** Potenza, Italy
**project** Urban renewal
**client** Municipality of Potenza
**structures** Favero&Milan Ingegneria
**systems** Favero&Milan Ingegneria
**plan** 2000
**construction** 2006-2010
**cost** € 2,066,737
**built area** 3,248 sq.m
**volume** 6,800 cu.m
**contractor** Giovanni Basentini Lavori S.r.l. – C.S.T. impianti S.r.l.

# 项目报告

## 隔音墙
**Impruneta，意大利，2000年-2007年**

此建筑和景观工程项目从属于一项针对A1高速公路佛罗伦萨至博洛尼亚段第3车道的高速公路基础设施综合性整治计划。

此项目最初的目的是计划和构建噪声防护带以及减少基础配套设施所带来的环境影响的解决方案，如交通枢纽停车场和人工地下通道。用赤陶和考登钢制作此隔音墙既具备强大的吸音能力，又具有考登钢和玻璃所带来的反射声波的能力，从设计方面来说，这种建筑构造能够让人联想到一种当地的传统材料，所以可为沿此高速路段的景观赋予一定的特色。此隔音墙采用一种微孔化考登钢面板系统制造，由"双T型"梁支撑，梁上有重复的赤陶要素线条序列装饰。在面板内加填矿棉隔音纤维能有效减低噪声影响，来自Impruneta镇的赤陶能够耐受烟雾和霜冻，从而大量降低了维护成本并赋予此隔音墙以古城墙般的外观效果。

**页号76-83**

**地点** 意大利佛罗伦萨Impruneta镇
**项目** 吸音和反射声波屏障
**客户** Autostrade per l'Italia S.p.A.
**结构** CIR-Ambiente
**规划** 2000年
**施工** 2005年-2007年
**成本** 6500000欧元
**建筑面积** 8,300,000平方米
**赤陶所覆盖表面积**
20000平方米
**承包商** CIR-ambiente

## VIA TIRRENO重建
**意大利，波坦察，2000年-2010年**

本项目是波坦察Cocuzzo区的工薪阶层的公寓楼所围合的一个空间，此空间处于两排建筑物之间，其附近包含了在很大程度上限制地面自由空间的街道，使得可用停车位处于最少水平，并完全忽略了植被和人行道存在。设计方案摒弃了只求美观的翻新方法，将这种环境背景当成一种矛盾来解决，这种矛盾存在于人工环境之中，即在横跨各建筑物外立面之间形成的这条"水泥峡谷"，与自然环境之间，在自然环境中，这两堵带着窗户墙壁之间的空间被视作一条深谷，一条谷底。此空间由砾石和岩石来定义，赋予了一个大型直道公园以生命力，降低了汽车和街道的影响。天然斜坡又构成了新的地表形式，让一部分隐入地下，将停车场隐蔽起来，形成了架高景观平台。在地面上，用摊铺绿色地面替换了沥青，并设置了用铁锈色染涂的立体水泥体块，给沿人行道的一系列街区带来了勃勃生机。

**页号84-95**

**地点** 意大利，Potenza
**项目** 城市更新
**客户** Potenza市政府
**结构** Favero&Milan Ingegneria
**系统** Favero&Milan Ingegneria
**规划** 2000年
**建设** 在建 2006年-2010年
**成本** 2066737欧元
**建筑面积** 3248平方米
**体量** 6800立方米
**承包商** Giovanni Basentini Lavori S.r.l.
– C.S.T. impianti S.r.l.

# ANTINORI WINERY
## San Casciano Val di Pesa, Italy, 2004

The site is surrounded by the unique hills of Chianti, covered with vineyards, half-way between Florence and Siena. A cultured and illuminated customer has made it possible to pursue, through architecture, the enhancement of the landscape and the surroundings as expression of the cultural and social valence of the place where wine is produced. The functional aspects have therefore become an essential part of a design itinerary which centres on the geo-morphological experimentation of a building understood as the most authentic expression of a desired symbiosis and merger between anthropic culture, the work of man, his work environment and the natural environment. The physical and intellectual construction of the winery pivots on the profound and deep-rooted ties with the land, a relationship which is so intense and suffered (also in terms of economic investment) as to make the architectural image conceal itself and blend into it.
The purpose of the project has therefore been to merge the building and the rural landscape; the industrial complex appears to be a part of the latter thanks to the roof, which has been turned into a plot of farmland cultivated with vines, interrupted, along the contour lines, by two horizontal cuts which let light into the interior and provide those inside the building with a view of the landscape through the imaginary construction of a diorama.
The façade, to use an expression typical of buildings, therefore extends horizontally along the natural slope, paced by the rows of vines which, along with the earth, form its "roof cover". The openings or cuts discreetly reveal the underground interior: the office areas, organized like a belvedere above the barricade, and the areas where the wine is produced are arranged along

the lower, and the bottling and storage areas along the upper. The secluded heart of the winery, where the wine matures in barrels, conveys, with its darkness and the rhythmic sequence of the terracotta vaults, the sacral dimension of a space which is hidden, not because of any desire to keep it out of sight but to guarantee the ideal thermo-hygrometric conditions for the slow maturing of the product. A reading of the architectural section of the building reveals that the altimetrical arrangement follows both the production process of the grapes which descend (as if by gravity) – from the point of arrival, to the fermentation silos to the underground barrel vault – and that of the visitors who on the contrary ascend from the parking area to the winery and the vineyards, through the production and display areas with the press, the area where vinsanto is aged, to finally reach the restaurant, the bakery and the floor hosting the auditorium, the museum, the library, the wine tasting areas and the sales outlet. The offices, the administrative areas and executive offices, located on the upper level, are paced by a sequence of internal court illuminated by circular holes scattered across the vineyard-roof. This system also serves to provide light for the guesthouse, the caretaker's dwelling and the employees' kindergarten. The materials and technologies evoke the local tradition with simplicity, coherently expressing the theme of studied naturalness, both in the use of terracotta and in the advisability of using the energy produced naturally by the earth to cool and insulate the winery, creating the ideal climatic conditions for the production of wine.
**pag 96-125**

**location** Bargino, San Casciano Val di Pesa, Florence, Italy
**project** Winery, offices
**client** Marchesi Antinori s.r.l.
**project management and engineering** Hydea s.r.l., Paolo Giustiniani
**structures** A&I progetti s.r.l., Massimo Toni, Niccolò De Robertis
**systems** M&E s.r.l., Stefano Mignani, Paolo Bonacorsi
**viniculture plant** Emex Engieneering S.r.l. Trading-pty-L.t.d.
**plan** 2004-2008
**construction** Under construction
**cost** € 65,000,000
**plot area** 139,950 sq.m
**built area** 41,165 sq.m
**volume** 287,260 cu.m
**contractor** Inso S.p.a.

# ANTINORI酿酒厂

**San Casciano Val di Pesa**，意大利，**2004**年

项目基地被基安蒂山脉环绕，有大片葡萄园种植基地，介于佛罗伦萨和锡耶纳的中间位置。具备深厚的文化背景的客户让我们能够通过建筑设计来增强建筑景观和周围环境的完美融合，表达出这个酿酒场所的文化和社会价值。 因此，功能变成了这个设计过程中最为核心的部分，我们集中于这样一座建筑物给人带来的地貌体验，让人理解到，这座建筑物就是人们所想达到的与自然共栖互利的最真实表达，并融人类文化、人的成就及其工作环境和自然环境于一体。这座酿酒厂的实体和精神建造以深植于这块土地的强烈联系为枢纽，项目投资巨大（从经济投资的角度），建筑形象与周边景观融合。这个项目的目标是将这座建筑物变成与乡间景观融合的巧夺天工的造物；这座工业性的综合建筑物从外观上成为景观的一部分。这种融合要归功于屋顶改造成了一种葡萄种植园。沿着等高线设置的的两道水平开口将连绵的屋顶种植园打断，将日光引入建筑内部。同时以虚构透视图的方式向屋内呈现观赏风景的角度。在立面上，采用这座建筑物典型的表达方式，沿着自然的坡度向水平方向做出延伸，间隔着一行行的葡萄藤，而这些葡萄藤与土地一起构成了"屋顶覆盖层"。这些开孔或开口隐约地揭示出地下的内饰：办公区组织得像地窖上的望景楼，用于生产葡萄酒的区域沿较低层排列，而灌装区和贮存区沿较高层布置。作为这座酿酒厂隐藏的核心区，也就是葡萄酒在木桶中酿成之处，以其幽暗且韵律分明的赤陶拱顶，传达出这个隐藏空间的神圣特征，不是因为不想让人们看到这个空间，而是为了保证这种葡萄酒酿造过程缓慢陈酿所需的理想温度与湿度条件。细品这座建筑物的各个部分就可察觉到，其高度上的安排既跟随着葡萄酒生产流程向低处走的特性（就像水自流一样），从到达点至发酵罐再到穹顶地窖，还跟随着来访者逆行向上的次序，从停车场至酿酒区再到葡萄园，经过设置了压榨机的生产和展示区，餐后酒的醇化区，并最终到达餐厅、面包房和布置着礼堂、博物馆、图书馆和品酒区以及销售区的楼层。办公室、行政管理区和经理办公室位于高层，间隔着一系列的内部庭院，这些庭院由散布于葡萄园式屋顶各处的圆形孔洞来提供自然照明。这个系统还用于为客房、门卫的居住处和雇员幼儿园照明。这个项目所采用的材料和技术让人联想起当地崇尚简约的传统，并清晰合理地表达了居者熟知自然状态的主题，这些同时体现在赤陶的运用，以及明智地运用土地天然产生的能量来冷却和保温这座酿酒厂的做法，为葡萄酒的生产创造了理想的气候条件。

页号96-125

**地点** Bargino, San Casciano Val di Pesa，佛罗伦萨，意大利
**项目** 酿酒厂、办公楼
**客户** Marchesi Antinori s.r.l.
项目管理和工程设计 Hydea s.r.l., Paolo Giustiniani
**结构设计** A&I progetti s.r.l., Massimo Toni, Niccolò De Robertis
**系统设计** M&E s.r.l., Stefano Mignani, Paolo Bonacorsi
**葡萄栽培装置** Emex Engieneering S.r.l. Trading-pty-L.t.d.
**规划** 2004年-2008年
**建设** 在建
**成本** 65,000,000欧元
**占地面积** 139,950平方米
**建筑面积** 41,165平方米
**体量** 287,260立方米
**承包商** Inso S.p.a.

# TIANJIN LAND 7 OFFICE BUILDING
**Tianjin, China 2005**

The project reorganizes part of a city block in the Italian district of Tianjin, built beginning in 1901 In one of the bends of the Haihe River. The request for a new office and commercial complex suggests a project that combines distributive necessities and contemporary functionality with the intention to preserve the look of the district organized on the reproposal of a minute urban fabric that confirms the typical traditional Italian city housing characteristics. Such a formulation rejects the advancement of macro-buildings, leftovers of an intensive culture and devoid of any recognizable character that could be distinguished from today's Chinese customs. The proposed solution is composed therefore as a combination of two contrasting models to enclose a central public space: the buildings with open-court plans, no more than four floors, search for a more direct dialogue with the scale of the adjacent houses, creating a system of internal squares and truncated perspectives constructed at pedestrian scale; on the opposite side a sort of horizontal unit, formalized in a long seven-floor block in terracotta colored concrete, encircling the block like a sort of urban wall fragment. The whole complex is crowned by roof gardens and crossed by porticoed passageways, optimization of public space use and of the ground level areas used for commercial functions.
**pag 126-135**

**location** Tianjin, China
**project** Management and commercial
**client** HEDO Construction
and Investment Group
**structures** Favero&Milan Ingegneria
**plan** 2005 competition, 2nd place
**cost** € 15,000,000
**built area** 18,000 sq.m
**volume** 90,000 cu.m

# MASTERPLAN OF CASTELLO
**Florence, Italy 2005-2007**

The project for the Castello plain sinks its roots into at least thirty years of Florentine history, evolving from a simple office district into a true portion of city.
The first hypotheses for a project date back to 1976 when Florence University announced a competition to come up with some ideas for a project for the office district, which was to characterise the northern part of the area. In 1978, with the second phase of the PIF (Florentine Inter-municipal Plan), new guidelines indicated the need to include more functions in the area (tertiary, managerial, cultural). In 1985, the adoption of the variant to the PRG (General Masterplan) denominated "north-west variant", stipulated 3 million cubic metres of building, as well as a park covering 60 hectares. In 1998, after years of debate and negotiating, a new variant to the PRG increased the park's dimensions to 80 hectares and drastically reduced the cubature to a total of 1,400,000 cubic metres. Subsequently, Fondiaria insurance company, owner of the majority of the land, agreed with the local authorities to employ Richard Rogers to redesign the entire area; in 1998, with Rogers' plan a new urban model was introduced, which rejected the zoning of the previous master plans, based on multifunction.
In 1999, on the PUE's approval of the local authorities' initiative, Rogers' project was confirmed.
Nevertheless, owing to contrasting political views, the project did not go ahead, and the local authorities, following negotiations in which Fondiaria ceded 24 hectares of land to the Municipality of Florence for the construction of the so-called "Scuola dei Marescialli" – a vast military police barracks – drew up and approved a new varient to

the PUE (Unitary Building Plan) in 2004. The authorities slightly increased the park's dimensions from 800,000 sq.m. to 805,000 sq.m, while the total cubature remained as was, including, however, also the public military part. In addition, a clause was added for flexibility (20% of useable surface area could be shifted from one lot to another). Hence in the new configuration, excluding the military areas already under construction, there was a theory that of the 160 remaining hectares, one half was destined for park land and the other half for building. In 2006, for the design of the park, Fondiaria's new owner called French landscape designer Christophe Girot, who teaches at the ETH school of Zurich, and also Studio Archea, in particular, Florentine-born Marco Casamonti, professor of architectural planning at Genoa University. The two studios, with the historical and critical advice of Vittorio Savi, embarked upon a complex study and research, establishing the general criteria of the new detailed master plan. The main paths of development involved the need to introduce environmental sustainability and varied architectural projects able to culturally define the intervention as a new part of the city. The main strategic elements are: functional mix (no longer dormitory or exclusively office and residential districts, but an area of shops, residences, offices and integrated public activities); architectural mix (detailed projects carried out by various architects, in line with important urban transformation projects carried out with this methodology throughout Europe); pedestrianisation of the area (through the availability of artificial terrain which contains road networks and car parks); integrated mobility (through the planned link of the area with a new tram

# 天津7号地块办公楼

天津, 中国, 2005年

# CASTELLO 总体规划

佛罗伦萨，意大利，2005年-2007年

此项目是对天津意大利风情区内的一个城市街区进行改建，此区始建于1901年，位于一个海河河弯处。根据这座新建办公和商业综合建筑的要求，本项目在布局上应和现代化的功能要求与保留街区面貌的意图结合起来，在组织上以重现细微城市生活基本结构为中心，确立典型的传统意大利城市房屋特性。这样的规划拒绝了宏大建筑物的激进手法，保留了一种风格强烈的文化，但避免任何可与当今中国风俗习惯明确区分出来的特色。所以建议的解决方案是将两种对比鲜明的模式结合起来，将一片中心公共场所环绕在内：这些建筑物采用了开放式庭院规划，楼高不超过4层，和周边建筑的体量相当，并在近人的尺度上建立内部广场空间和观景点间的关系；对面一侧采用一种水平单元，以赤陶色混凝土构成的较长7层建筑，像城墙一样围绕着这个街区。整个综合建筑物屋顶覆屋顶花园并有若干带门廊的走道穿越其中，优化了公共场所利用以及用于商业功能的地面层区域。

**页号126-135**

**地点**　中国天津
**项目**　管理和商业
**客户**　HEDO Construction and Investment Group
**结构**　Favero&Milan Ingegneria
**规划**　2005 年竞赛，第2名
**成本**　15,000,000欧元
**建筑面积**　18,000平方米
**体量**　90,000立方米

这个位于Castello平原的项目在佛罗伦萨至少有30年的历史，从一座简单的办公区演化成真正属于这座城市的部分。第一次的项目假想可追溯至1976年，当时佛罗伦萨大学宣布了一项竞赛，要求大家为这个办公区项目提出一些创意，用于突出这片区域北部的特色。在1978年，在PIF（佛罗伦萨城市间合作计划）的第二阶段，新的指导方针表明了需要在这个区域包含更多的职能（第三级、管理、文化）。1985年，采纳了PRG（总体规划）的改良方案并命名为"西北方案"，其中规定了300万立方米的建筑体量以及一座占地60英亩的公园。1998年，在多年辩论和谈判之后，PRG新改良方案将公园的面积增加到80英亩，并极大幅度地削减体量至1,400,000立方米。随后，这块土地的主要业主Fondiaria保险公司与当地政府达成协议，聘请Richard Rogers来重新设计整个区域；1998年，根据Rogers的规划引入一个新城市模型，回避了之前总体规则以多功能为基础的分区设计方式。1999年，根据当地政府的PUE批复，确认了这个"initiative, Rogers"项目。尽管如此，由于相互对立的政见，这个项目没有能够继续下去，经过谈判，Fondiaria保险公司将其中24英亩土地退还给佛罗伦萨市政府，用于建设称为"Scuola dei Marescialli"这一座占地庞大的宪兵兵营，当地政府于2004年起草并批准了PUE（统一建设计划）的一个新改良方案。当地政府略微增加了公园的规模，从800,000平方米增至805,000平方米，而总体体量仍然保持如前，同时还包括了军事设施部分。此外，为了增加灵活性还增设了一个条款（20%的可用面积可以从一个场地转移给另一个场地）。因此，在新配置方案中，除了已经在建的军事区域以外，在理论上，这剩余的160英亩土地中，一半预定用作公园，另一半用于建筑物。2006年，Fondiaria的新业主邀请法式园林设计师,在苏黎世ETH学校执教的Christophe Girot及阿克雅建筑事务所，尤其是事务所创办人之一热那亚大学建筑规划专业教授佛罗伦萨人Marco Casamonti在Vittorio Savi的关键性历史学究性建议下，展开了一项复杂的研究和探索。确立了新的详细总体规则的总体标准。涉及了引入环境可持续发展的要求以及让各种建筑能够从文化上将这个项目纳为城市新的组成部分。主要策略是：功能上混合（不再是宿舍区或只是办公楼和居民区，而是一块拥有商店、住宅、办公楼和一体化公共活动的区域）；建筑设计上混合（由各种类型的建筑师来执行具体项目，与在欧洲各地的重要城市改造项目所采用的方法保持一致）；区域内的行道（通过规划形成路网和停车场）；一体化出行体系（通过规划将本区域与一条新建有轨电车线路连接，从而与佛罗伦萨市中心连接起来）；重要公共职能（将省政府和区政府办公楼迁移至城市的入口，靠近高速公路和机场）；公共地域塑造（通过街区密切的分布与宜人的尺度来设计，完全由公众使用）；将公园作为结构要素（此处的绿化将不仅简单地认为是附属元素而是这个四通八达的城市新基本结构的鲜明特色）。

从建筑设计的角度来看，这个项目建立了以建筑物边缘为参照的过渡体系：面向城市和街道的立面稳固而坚硬；另一方面，面向公园的一面，建筑物是开放的，偏构筑空间与绿色空间之间互相渗透。

line connecting the area of Castello with the centre of Florence); important public functions (with the plan of shifting the offices of the Province and Region to the gates of the city, near the motorway and airport); public use of the terrain (through the articulated people-friendly distribution of surfaces to develop, completely for public use); the park as a structural element (where greenery is not simply considered an accessory element but a distinctive characteristic of this new part of interconnected urban fabric).

From an architectural point of view, the project establishes a hierarchy in relation to the limits of the building: toward the city and street, the façades are solid and rigid; in the park, on the other hand, the buildings open up, favouring co-penetration between the constructed space and green space.

The urban fabric in the project appears in continuity with the consolidated city, constantly evolving among a series of squares which cross a pedestrian main street at regular intervals, surrounded by porticoes and shops, culminating in the square where the Regional and Provincial Offices are to be located.

**pag 136-145**

**location** Castello, Florence, Italy
**project** Residential, commercial, directional
**client** Europrogetti S.r.l.
**plan** 2005-2007
**landscape project** Atelier Girot
**plot area** 120 ha
**park area** 80 ha

# REGIONAL GOVERNMENT BUILDING
**Florence, Italy, 2005-2007**

The project regards a building for public management, designed as a Headquarters for the Tuscan Region. The architectural complex develops on 6 floors with a total surface area of 68,108 sq.m The project rejects the hypothesis of a single building and instead develops on an urban scale, resulting in a sort of city block formed by 7 building elements, linked by a functional basement in turn placed upon a slab where 2 levels of car parks are located. The project, consequently, represents the design of a portion of city, where the spaces appear sealed among the buildings, or inside private courtyards, while the rapport between the building and public road remains contiguous according to the example of consolidated urban structures. The ground floor of the complex is a large space linking the elements overlooking it, occupied by conference halls, an auditorium, a library, a canteen, nursery, gymnasium and in general all the common services that can be directly accessed by the public. Two large tree-lined courtyards are lit up. The entrance to the building may be accessed on foot both via the basement on the south-west and south-east sides, and via the raised square, through openings on the individual building elements. The various buildings, clad in large stone slabs of various hues, are characterised by pitched roofs clad in the same material.

**pag 146-153**

**location** Castello, Florence, Italy
**project** Offices
**client** Europrogetti S.r.l.
**systems** Hilson&Moran
**plan** 2005-2007
**cost** € 150,000,000
**built area** 68,000 sq.m
**volume** 238,000 cu.m

# DISTRICT GOVERNMENT BUILDING
**Florence, Italy, 2005-2007**

The project involves a building for "private management", nonetheless designed as a Headquarters for the Province of Florence. The architectural complex develops on 6 floors with a total surface area of 20,247 sq.m. The new building stands within a lot as the conclusive part of a complex urban system, so as to create a sequence of public spaces circumscribed by two elements of building immersed in greenery, also containing a private raised square, also for public use. This is linked, through means of a great flight of steps, to the true public space, which converges in the "main street", characterising and crossing the entire masterplan of Castello.

The public, pedestrianised spaces are located in a privileged position in relation to the surroundings, since organised upon a large podium which provides an observation point with a view over the park. The decision to build on a raised level furthermore resolves the risk of flooding.

From a point of view of form, the building configures as a sequence of slightly staggered floors, which revives the theme of terracing so typical of the Tuscan hills. The entirely glazed façades contain a copper micro-tissue towards the exterior and an iconographic film towards the internal square, calling to mind the pictorial scenes of Florence in the Salone dei Cinquecento.

**pag 154-161**

**location** Castello, Florence, Italy
**project** Offices
**client** Europrogetti S.r.l.
**systems** Hilson&Moran
**plan** 2005-2007
**cost** € 50,000,000
**built area** 20,000 sq.m
**volume** 70,000 cu.m

项目中的城市规划肌理看起来与城市融为了一体，沿着一系列的广场不断演进着，以规则的间隔穿越行人主道，周边环绕着柱廊和商店，直至座落大区与省政府办公楼的制高点上的广场。

**页号136-145**

**地点**　Castello, Florence, Italy
**项目**　Residential, commercial, directional
**客户**　Europrogetti S.r.l.
**规划**　2005年-2007年
**景观项目**　Atelier Girot
**占地面积**　1,200,000平方米
**公园面积**　800,000平方米

## 托斯卡纳大区政府办公楼
### 佛罗伦萨，意大利，2005年-2007年

此项目是一座公共管理用的建筑物，设计作为托斯卡纳大区（Tuscan Region）的总部。这一建筑群在6层楼基础上开发，总面积68108平方米。项目舍弃了单幢建筑物的形式，而是按城市规模来开发，最终形成一由7项建筑物元素构成的城市街区，由功能性地下室连接，下置混凝土楼板，地下室的功能为2层停车场。因此，这个项目就表达出了对一个城市组成部分的设计理念，各个空间看起来密封于各幢建筑物之间，或身处私密庭院之内，按照稳固的城市结构，建筑物和公共道路之间保持融洽关系。这座建筑群的底层是一个大型空间，连接着各个空间要素，这里布置了会议厅、礼堂、图书馆、食堂、托儿所、健身房以及一般性的公众可以直接享受的所有服务。两块大型林荫庭院点亮了这个空间。这座建筑物的入口可以徒步到达，既可经过位于西南侧和东南侧的地下室，又可经过架高广场，达到各个建筑空间的入口处。各种不同建筑物覆盖着不同色调的大片石材，并以覆盖着相同材质的倾斜屋顶作为共同特征。

**页号146-153**

**地点**　Castello，佛罗伦萨，意大利
**项目**　办公楼
**客户**　Europrogetti S.r.l.
**系统**　Hilson&Moran
**规划**　2005年-2007年
**成本**　150,000,000欧元
**建筑面积**　68000平方米
**体量**　238,000立方米

## 佛罗伦萨省政府大楼
### 佛罗伦萨，意大利，2005年-2007年

此项目涉及到一座用于"私营企业管理"的建筑物，虽然设计作为佛罗伦萨省的总部。这座建筑群以6层楼的形式开发，总面积20247平方米。这座新建筑物所坐落的地块是一个复杂城市系统的最后组成部分，并形成了一连串的公共空间，掩映于草木之间由两个建筑单元环抱一个抬高的作为公用的内庭广场。这个广场通过大段的阶梯与真正的公众空间相连，而这个真正的公众空间又融合到了"大街"之内，这条大街穿越了整个Castello的规划布局并形成了这个规划的特色。

这些步行的公众空间拥有一种优于周围环境的地位，这是因为它们设置在一个抬高的平台上，而平台为人们提供了一个可以俯瞰公园的观景点。抬高高度的建设决策还解决了遭遇洪水的危险。

从形式来看，建筑物配置成了一连串略有交错的楼层，让人想起托斯卡纳山岳典型的山丘形象。双层玻璃幕墙的内层为夹设铜制网片的夹层玻璃，面向广场的外层同为夹层玻璃，夹置描绘佛罗伦萨老宫五百人厅内诗歌性油画的胶片。

**页号154-161**

**地点**　Castello，佛罗伦萨，意大利
**项目**　办公楼
**客户**　Europrogetti S.r.l.
**系统**　Hilson&Moran
**规划**　2005年-2007年
**成本**　50,000,000欧元
**建筑面积**　20,000平方米
**体量**　70,000立方米

## INFO POINT
**Florence, Italy, 2005-2007**

The project involves the construction of a temporary structure to be used on a building sitc and also as an information and meeting point for citizens who wish to keep up to date with the various transformations occurring in the area of Castello.

The structure not only fulfils the task of divulging the building methods in the area, but also takes on the symbolic value of gateway to the new project, being situated between the main access road, with one part within the park land and the other in the middle of the roundabout.

The object in question therefore takes on powerful representative value, both in terms of form and of content and ability to narrate, since it represents itself as a kind of manifesto for architecture that can make a large surface habitable without consuming the terrain. In fact, although the building covers an overall area of some 1,500 sq.m, it occupies ground space of just one tenth of the surface occupied, and hence taken from the park. Also foreseen is an attentive plant engineering project, which, although temporary, flanks its performance with highly efficient bioclimatic buildings.

**pag 162-169**

**location** Castello, Florence, Italy
**project** Offices
**client** Europrogetti S.r.l.
**structures** Favero&Milan Ingegneria
**systems** Hilson&Moran
**plan** 2005-2007
**cost** € 5,000,000
**built area** 1.000 sq.m

## HIGHWAY MUSEUM
**Salerno-Reggio Calabria, Italy, 2006**

The competition centres on an area by the 41$^{st}$ kilometre of the A3 highway which is to be landscaped and where a museum centre is to be created, to diffuse and provide information on the archaeological findings. The proposal consists of a large disc with a diameter of more than 230 metres, suspended on the highway, conceived as a hybrid between new ground and roof. A flat, quite singular building, at the same time landmark, inhabited bridge allowing people to cross the road, and rest and stop for a while, and a centre for the production of renewable energies. The disc defines an intangible physical environment where paths and spaces associated with the three themes of the exhibitions (archaeology, nature, history of the motorway) are intertwined among them and with the entertainment and leisure activities required. The volumes above ground and the service and rest areas are designed as an extension of the natural elements, while the open space is conceived as a landscape modelled by the architectural elements: the ground is developed in rises and depressions which describe gardens and paths, to then dissolve and merge with the surroundings.

**pag 170-175**

**with Pietro Carlo Pellegrini**
**and Studio Franchi Lunardini Partners**
**location** Highway between Salerno and Reggio Calabria, Italy
**project** Landscape restoring and highway museum
**client** ANAS S.p.A.
**structures** Giuliano Sauli
**systems** Sistemi Industriali S.r.l.
**plan** 2006 competition, 1$^{st}$ place
**cost** € 20,000,000
**green area** 15,000 sq.m
**parking area** 55,820 sq.m

## TICOSA EX INDUSTRIAL AREA
**Como, Italy, 2006-2007**

The area created by the demolition of the Ticosa textile plant lies along the directrix of access to the old town of Como. The master plan features the creation of dwellings, shops, public services and gardens, defining the new viability below an artificial ground, cultivated as a park, eliminating all vehicle traffic "on the surface". The project reconnects the most important buildings in the urban tissue (the convent, the basilica, the old town, the cemetery) which have been overshadowed by the industrial past of the area, assuming the role of strategic junction connecting the "high" level of the cemetery with the level of the city. The proposal defines a system of relations and functional integrations with a vertical development which makes it possible to arrange the buildings on the altitude of the public park. The difference in altitude has been concealed by a complex system of stairs and paths which connect the shops on the same level as the city with services and the park. The latter, a mediation between monumental cemetery and new district, serves as connection to the square, the green strip and the areas covered by trees, providing a continuity of the urban tissue to the new residences. A single tall building, with 13 floors, performs the role of vertical reference among the profiles proposed by the project.

**pag 176-187**

**location** Areas of the former Ticosa textile industry, Como, Italy
**project** Residential and commercial
**client** Multi Veste Italy S.r.l.
**structures** BMS Progetti
**systems** Studio Proima
**plan** 2006-2007
**cost** € 60,000,000
**green area** 15,000 sq.m
**parking area** 55,820 sq.m
**built area** 39,649 sq.m

# 信息站

**佛罗伦萨，意大利，2005年-2007年**

此项目是建设在一个项目施工现场上的临时性构筑物，而且作为市民们的一个信息交流和会议地点，提供他们希望跟踪了解Castello区正在发生的各种改造项目情况的信息需要。

这个构筑物不仅要提供本地区建设进展的信息，还要承担作为通往新项目大门的标志，其位置部分跨越主要入口道路，部分处于公园土地上，另一部分位于环行交叉路的中间。

因此，我们所要谈及的对象承载了强大的代表性作用，既涉及到代表的形式和内容，也涉及到叙述的能力，因为它本身代表着这样一种建筑设计宣言，我们能够在不破坏地形的条件下建设大量居住面积。事实上，尽管这座建筑总面积大约有1500平方米，但其占地面积仅为建筑面积的1／10，而土地用以建设公园。而且可以预见它还是一项在设备工程设计上体贴入微的项目，尽管属于临时性构筑物，仍然以高效的生态建筑来体现了它的性能。

**页号162-169**

| | |
|---|---|
| **地点** | Castello, Florence, Italy |
| **项目** | 办公楼 |
| **客户** | Europrogetti S.r.l. |
| **结构** | Favero&Milan Ingegneria |
| **系统** | Hilson&Moran |
| **规划** | 2005年-2007年 |
| **成本** | 5,000,000欧元 |
| **建筑面积** | 1,000平方米 |

# 高速公路博物馆

**Salerno-Reggio Calabria，意大利，2006年**

此竞赛项目的中心是一块位于A3高速公路第41千米旁的一个区域，这个区域将进行景观整治并将建立一个博物馆中心，用于传播和提供有关考古学发现的信息。设计由一个直径超过230米的大型碟形建筑构成，它悬挂在高速公路上方，构想成为新地面与屋顶之间的混合体。这是一个外形扁平但极为特异的建筑物，同时又是地标建筑物，内含让人们可以跨越道路的桥梁，并且可供人们停留和休息一阵子，并设有再生能源生产中心。这个碟形建筑定义了一种本来难以确定的实体环境，让路径和空间与3个展览主题（考古学、自然、高速公路历史）关联起来，并相互之间盘旋缠绕，并提供了所需求的娱乐和休闲活动空间。地面上的体量和服务休息区设计为天然元素的延伸，而开放的空间又是由各种建筑设计元素塑造的景观：地面层在建设中形成了高地和洼地，构成了花园和道路，然后溶解与合并至周围环境之中。

**页号170-175**

**与Pietro Carlo Pellegrini和Studio Franchi Lunardini Partners合作**

| | |
|---|---|
| **地点** | 萨勒诺与勒佐卡拉布里亚间高速公路，意大利 |
| **项目** | 景观恢复和高速公路博物馆 |
| **客户** | ANAS S.p.A. |
| **结构** | Giuliano Sauli |
| **系统** | Sistemi Industriali S.r.l. |
| **规划** | 2006年设计竞赛，第一名 |
| **成本** | 20,000,000欧元 |
| **绿地** | 15,000平方米 |
| **停车面积** | 55,820平方米 |

# TICOSA EX工业区

**科摩，意大利，2005年-2007年**

此区域由Ticosa纺织厂拆除后形成，位于通向科摩旧城道路准线旁。总体规划包括了建设居住区、商店、公共服务设施和公园，设计赋予了在人造地面下的新生活力，设置乐停车场，消除"地面上"的一切车辆交通。项目重新连接起城市肌理中的各个最重要建筑物（修道院、大教堂、旧城、公墓），而这些建筑物都曾被这个地区过去的工业形象所遮蔽；项目还承担着连接公墓"高"标高与城市标高之间战略性结点的功能。新的设计定义了一个关系体系和功能性集成设施，采用了垂直发展方式，使得建筑物可以安排在起伏公园的不同标高上。高度上的差异的隐藏由一套连接着与城市相同标高的商店以及服务设施和公园的复杂的阶梯系统和道路来实现。道路作为纪念公墓与新地区之间的沟通，成为了广场、绿化带和树林笼罩区域的连接枢纽，在城市肌理与新居民区之间提供了连续性。其中一栋13层的单幢建筑成为了项目中所提出的各单体垂直参照物。

**页号176-187**

| | |
|---|---|
| **地点** | 前Ticosa纺织厂区域，科摩，意大利 |
| **项目** | 住宅和商业 |
| **客户** | Multi Veste Italy S.r.l. |
| **结构** | BMS Progetti |
| **系统** | Studio Proima |
| **规划** | 2006年 -2007年 |
| **成本** | 60,000,000欧元 |
| **绿地** | 15,000平方米 |
| **停车面积** | 55,820平方米 |
| **建筑面积** | 39,649平方米 |

# BORGO ARNOLFO
**San Giovanni Valdarno, Italy, 2006**

The former hospital of San Giovanni Valdarno faces the main thoroughfare to the historic center and marks its entrance point. One of the projects part of the city's strategic plan, following the construction of a new hospital designed by Vittorio Gregotti, the conversion for residential and commercial use of the area entailed demolishing the old building and reconfiguring the spaces surrounding it. The design envisioned an urban model based on the principles of closing and bounding while simultaneously opening and making accessible. This concept is epitomized in the architectural conception of the square. Through carefully planned openings, the design gives a new visual perspective on the pedestrian thoroughfare, which leads to the historic center, and of which the project is a natural extension. A large projecting portico extends along all sides of the buildings' perimeter. The façades call a city block to mind with the sight of a sequence of urban facades, with the staggered layout and varying heights of contiguous structures of a varied sequence of buildings and roofs, moving up to the top of a tower, in a medieval type organization of space.
**pag 188-197**

**location** San Giovanni Valdarno, Florence, Italy
**project** Residential, commercial
**client** Etruria Investimenti S.p.A.
**structures** Studio Bacci & Bandini ingegneri associati
**systems** Giuliano Galzigni
**acoustic** Annalisa Baracchi
**plan** 2006-2008
**construction** Under construction
**cost** € 20,000,000
**area** 7,686 sq.m
**volume** 25,365 cu.m
**contractor** Etruria Investimenti S.c.a.r.l.

# ALBATROS CAMPING
**San Vincenzo, Italy, 2006-2009**

The project is aimed at the architectural and environmental upgrading of local infrastructures for tourists with a limited budget. The project is inspired by the conformation of marine organisms; the area has been arranged in a system of nuclei organized in two environmental macro-areas. The access area consists of a nature park, characterized by the spiral plan of the footpaths and bicycle tracks, which lead to the sports facilities. The ground has been remodelled in circular extrusions: two "filled" ones and one which has been excavated to create a natural open-air theatre. The camping area features a cluster of shops, cafes and entertainment venues organized in two theme areas distributing squares and services in concentric strips with paths that wind along the perimeter in a radial arrangement. The technical-constructive aspect aims to make the new service buildings blend into the environment, to cancel the impact of the buildings in the landscape. Every new construction is covered by straw and wood to create an environment which merges with the surrounding nature.
**pag 198-219**

**location** San Vincenzo, Livorno, Italy
**project** Hospitality
**client** Park Albatros s.a.s.
**structures** Roberto Nocentini
**systems** Leonardo Bracciali
**electric systems** S.T.E. società toscana elettrica S.r.l.
**pool systems** Acqua Sport Service S.r.l.
**plan** 2006
**construction** 2007-2009
**cost** € 20,000,000
**plot area** 40 ha
**sanitary blocks area** 1,100 sq.m
**market area** 1,650 sq.m
**contractor** T.I.S.

# PARKOUR - BEIJING
**Beijing, China, 2007**

Beijing Parkour represents a new housing strategy proposed as a reflection on the project in the historic Peking building holdings characterized by the presence of the typical Hutong. The project is interpreted as the idea of multiplication and recovery of area in a way that allows a more intensive development of these traditional areas without negating the typical horizontal course that distinguishes the oldest area of the city. The definition itself of Parkour, as "the art of knowing onself moving" explains the design approach: transience as social vehicle, changing the living space without altering the memory destroyed by the adherence to intensive occidental models shown in the vertical development of projects. A subterranean area connected to the existing structure frees Hutong's labyrinthine web into the third dimension allowing the elimination of spatial barriers and of the siheyuan (traditional courtyard houses) fencing, while retaining the proportions and the morphological structure. The spatial continuity overcomes the distinction between private and public, and opens it to various use possibilities: residential, commercial, cultural. A combination of prefabricated elements integrates the existing building fabric in a three-dimensional complex of free floors, served by escalators and equipped with service areas. The permeability of the system allows, via vehicular and pedestrian pathways, the penetration of the urban fabric at various additional levels optimizing vehicular circulation, usually problematic in historic districts.
**pag 220-229**

**location** Beijing, China
**project** Residential and commercial
**client** Beijing Xisi-Bei Ltd.
**plan** 2007
**built area** 18,000 sq.m
**volume** 200,000 cu.m

# BORGO ARNOLFO

San Giovanni Valdarno市，意大利，2006年

San Giovanni Valdarno市的这座前医院面对着通向具有历史意义的中心区的大道并标志着这个中心区的进入点。这座城市的战略规划也属于本项目的一个组成部分，在建设好由Vittorio Gregotti设计的新医院后，这块区域将改造为住宅和商业用途，就有必要拆除旧建筑物并重新配置围绕着此区域的各个空间。这个设计方案塑造了一种边界的封闭性和开放的可达性的完美融合模式。这个概念浓缩在广场的建筑设计构思之内。通过精心规划的开口，这个设计方案在主干道上给出新的视觉景观，这一景观将人们引向富有历史意义的市中心，使得本项目成为此中心的自然延伸。沿此建筑周边所有方向均设计了大型的外突柱廊。这些外立面设计，让这片街区在交错式布局和多变建筑及屋顶序列的连续结构和高低错落的引导下，形成了一种类中世纪的空间组织，为人民呈现了一系列丰富的城市街景印象。

外突出的柱廊。这些正面让一个城市区块唤起人们注意到一连串的城市正面景象，在交错式布局和多变建筑物和屋顶序列的连续结构高低错落的引导下，人们的视线会一直向上延伸到一座塔楼的楼顶，形成一种中世纪类型的空间组织。

**页号188-197**

**地点**　San Giovanni Valdarno，佛罗伦萨，意大利
**项目**　住宅楼，商用楼
**客户**　Etruria Investimenti S.p.A.
**结构**　Studio Bacci & Bandini ingegneri associati
**系统**　Giuliano Galzigni
**隔音**　Annalisa Baracchi
**规划**　2006年-2008年
**建设**　在建
**成本**　20,000,000欧元
**面积**　7,686平方米
**体量**　25,365立方米
**承包商**　Etruria Investimenti S.c.a.r.l.

# ALBATROS宿营地

San Vincenzo，意大利，2006年-2009年

本项目的目标是在有限预算内对当地供旅游者使用的基础设施进行扩建和环境优化。项目的灵感来自于海洋生物的形态；这块区域布置成由两个大环境核心体系，入口由一个自然公园构成，其特色体现在人行小道和自行车小径的螺旋式平面布局，并引向体育设施。地面经改造成为了圆形挤压式构造：两个"填造"的构造和一个开挖形成天然露天剧院的构造。宿营区的特色体现在一组商店、咖啡馆和娱乐场所，这些场所组织成两个主题区域，沿同心轴分布着各个广场和服务设施，其中还有小路按径向布置方式蜿蜒在圆周周边。方案在技术建设性方面的目的是让新建服务建筑物融入到环境之中，消除建筑物对景观的冲击。每一座新建构筑物都覆盖着稻草和木头，形成了一种与四围自然融合的环境。

**页号198-219**

**地点**　San Vincenzo，里窝那，意大利
**项目**　酒店设施
**客户**　Park Albatros s.a.s.
**结构**　Roberto Nocentini
**系统**　Leonardo Bracciali
**电气系统**　S.T.E. società toscana elettrica S.r.l.
**泳池系统**　Acqua Sport Service S.r.l.
**规划**　2006年
**建设**　2007年-2009年
**成本**　20,000,000欧元
**占地面积**　40公顷
**公共卫生区块面积**　1,100平方米
**市场面积**　1,650平方米
**承包商**　T.I.S.

# 跑酷（PARKOUR）– 北京

北京，中国，2007年

北京跑酷（Beijing Parkour）代表着一种新提出的房屋建筑策略，本项目身处于典型胡同建筑环境之中。项目可以解读为对这种区域进行增殖和恢复的设计思路，所采用的方法要让这些传统区域能够更加深入地发展下去，并区别于以往那些传统老区的发展水平面。跑酷（Parkour）本身的定义"知道自己在移动的艺术"解释了这种设计方法：作为社会载体的短暂性，改变生活空间而不改变那些被项目垂直方向开发所显示的西方模式毁掉胡同的空间记忆。一个连接至现有构筑物的地表下区域将胡同迷宫似的网络解放出来，使其进入三维世界，消除了空间障碍和四合院（传统庭院式房屋）的蔷墙，同时保留了比例和形态结构。空间上的连续性克服了私有空间与公用空间之间的严格界限，开发出各种空间的可能性：住宅、商用、文化。各种预制元素与现有建筑基本结构的结合形成了一个三维尺度的建筑物，由多个自由通行的楼层组成，配备了自动扶梯和设置了服务区。这个体系可允许这种城市基本结构通过车辆和行人通道在各个层面进行渗透，优化了车辆通行能力，而这种能力通常是历史性街区的问题所在。

**页号220-229**

**地点**　北京，中国
**项目**　住宅和商业
**客户**　Beijing Xisi-Bei Ltd.
**规划**　2007年
**建筑面积**　18,000平方米
**体量**　200,000立方米

## SHANGRI-LA WINERY
**Penglai, China, 2007**

The site selected for the location of the Shangri-la complex in Penglai extends in the north-south direction, between the highway route and the coastline: around forty hectares, ten of which are assigned to the building of the wine cellar and of a resort. The site morphology is characterized by a hillside which slopes from the road toward the sea; the reevaluation of the landscape leaves a mark on the design concept, based on the idea of a new scenario that integrates nature and artifice. The evocative image of the lighthouse – modernized to create a "fluctuating tower" that dominates the landscape – is accepted as a symbolic reference and as the recognition of the site's identity, for the visitors and motorists passing through. The underground cellar area is laid out in a north-south direction, perpendicular to the slope, creating a new artificial landscape. The new colored cement façade acts as a natural light filter and regulator for the production space. The same pattern is repeated in the landscape and in the building cladding above ground. The tower space houses the functions related to visitors, offices, exhibition area, and sales. This volume is also characterized by a horizontal notch that becomes the privileged vantage point of the entire landscape.
**pag 230-239**

**location** Penglai, China
**project** Industrial
**client** Shangri-la Winery Ltd.
**plan** 2007
**plot area** 8 ha
**built area** 18,000 sq.m
**volume** 72,000 cu.m

## EX CEMENTIFICIO
**Incisa Valdarno, Italy, 2008**

The site of the former Italcementi cement works is a link point between the historic center's urban fabric and its modern expansions. It fills in the area between the hill and the Arno River as a new urban center defined by service industry, administration and residential centers.
In terms of urban planning, the area is a piece of the reconfiguration of the Valdarno's landscape. The site's complexity, the lay of its land, the vicinity to the Arno and being crossed by the Borro della Fornacina channel all led to a number of considerations meant to place the architecture in an organic relationship with the site. The roofs are green and connect with the land to form new contours, an artificial land configuration made of walled terraces. The design of the public space and its two squares designates ground-level spaces for small retail businesses and public services, extending the natural pedestrian thoroughfare to the historic center; the role of the rampart is important as a mark remembering the historic manufacturing business. The volume covers from the 4th to the 6/7th floor with a series of terraces that are arranged in a curved staggered configuration that shapes a new artificial landscape next to the hill.
**pag 240-247**

**location** Incisa Valdarno, Florence, Italy
**project** Residential and commercial
**client** Siena Est S.r.l
**structures** Area Engineering
**systems** Clanis Progetti
**plan** 2008
**construction** 2010
**cost €** 15,000,000
**site area** 18,000 sq.m
**built area** 6,850 sq.m
**volume** 20,635 cu.m
**contractor** Siena Est S.r.l

## MULTIPURPOSE AND COMMERCIAL CENTRE
**Tirana, Albania, 2008**

The area of Quemala Stafa Stadium is part of the area of a plan aimed at the construction of a representative image of Tirana as a contemporary European capital. The design hypothesis relative to the south sector of the competition site represents instances of urban renewal implemented with the determination of mayor Edy Rama: "experimentation, color, nature". Formula that replaces the old slogan, applying it to the rehabilitation of monuments and historical façades, anonymous building additions of the '60s and '70s, and public and green spaces.
Three volumes – between 40 and 60 meters in height – emphasize the reference to the traditional tectonic in the gray stone block covering, broken up by narrow double height openings which accentuate the vertical development. The base of the buildings become instead a metaphor of contemporary Tirana: areas of red ceramic cut across the volumes according to the different surfaces which reveal the chromatic essence and the destruction of stereometric rigor in an arranged series of public and passage spaces covered for shopping and meeting. On the north side of the site the landscape is designed according to a geometric texture of green areas designed through a gradual expansion towards the city center. Parks and public spaces are part of a single system of urban routes, which nullify the presence of buildings as obstacles to the social use of public areas.
**pag 248-259**

**location** Tirana, Albania
**project** Commercial
**client** City of Tirana - ManeTCI
**plan** 2009 competition, 2nd place

# 香格里拉酒庄
蓬莱，中国，2007 年

蓬莱香格里拉建筑群的建在高速公路道路与海岸线之间向南北方向的延伸区域。面积大约40公顷，其中10公顷指定用于酒窖和渡假建筑。这个建址的地形特色体现在它毗邻一个山坡，这个山坡从道路向海的方向倾斜；设计师对景观的重新评估为设计留下了非常重要的一笔，并将自然和巧妙设计融合成新场景的创意为基础。灯塔拥有一种容易使人产生联想的形象，经现代建模化之后形成一种主宰整个景观"波动式灯塔"景象，对于参观者和经过的乘车者来说，这个景象成为他们身处区位的强大参照物，并作为此区域地标性建筑矗立。地下酒窖区向南北方向布置，与斜坡成正交方向，形成一种新的人造景观。新型彩色混凝土外立面对生产空间起到了自然光过滤和调节的作用。同样的设计方式也在景观当中以及地面覆盖建筑物内出现。灯塔内容纳着各种与观光相关的功能区、办公室、展示区和销售区。这个塔楼的另一大特色就是内含一层360度的酒庄最佳室外观景平台。

**页码230-239**

**地点** 蓬莱，中国
**项目** 工业
**客户** Shangri-la Winery Ltd.
**规划** 2007 年
**占地面积** 80,000平方米
**建筑面积** 18,000平方米
**体量** 72,000立方米

# 前水泥厂改造
Incisa Valdarno，意大利，2008 年

这个前Italcementi水泥厂建址位于其古城中心和现代城扩建部分之间的交接区。它座落于山丘与Arno河之间，成为一个新的城市中心，定义了服务行业、行政管理和居住中心。
从城市规划的方面来看，这个区域是对Valdarno市景观的重新配置。这个建址的复杂性，其地貌与Arno河的邻近关系以及穿越其间的Borro della Fornacina水渠，设计师都对其进行了一系列的设计思考，将建筑设计置于与这块建址的有机联系之中。植被绿地于大地衔接，构成新的等高线，一种由建有扶壁的梯田形成人造地形。公共空间及其两个广场的设计规定了地面层空间用于小型零售商业和公共服务，将自然的行人大道延伸至古城中心；城墙作为回忆有历史意义的制造业的标记物。项目第4层至6/7层楼设置一系列露台，采用了曲线形的交错配置形式，形成了一个紧挨山丘之后的新人造景观。

**页号 240-247**

**地点** Incisa Valdarno，佛罗伦萨，意大利
**项目** 住宅和商业
**客户** Siena Est S.r.l
**结构** Area Engineering
**系统** Clanis Progetti
**规划** 2008年
**建设** 2010年
**成本** 15,000,000欧元
**总平面面积** 18,000平方米
**建筑面积** 6,850平方米
**体量** 20,635立方米
**承包商** Siena Est S.r.l

# 多功能商业中心
地拉那，阿尔巴尼亚，2008 年

Quemala Stafa体育馆区域是一项旨在将地拉那市建设成为当代欧洲首都之一的规划的组成部分。
设计方案前提与此次设计竞赛建址的南部段相关，代表着在Edy Rama市长坚定决心下所实现的城市更新工程的实例："试验、色彩、自然"。用程序来替代旧口号，将程序运用于纪念性建筑和历史建筑外立面，无名的60、70年代建筑物、公共空间和绿地。
3种体量，高度在40米至60米之间，强调了对灰色石头街区屋顶所体现的传统建筑特色的借鉴，并以狭窄的双层通高开口作为中断以突出垂直发展的样式。建筑物的基础成了当代地拉那的隐喻：不同面横切建筑体块的表面，切面为鲜红的陶砖镶面，打破了建筑严谨的体量，作为商业零售和会议使用。在建址的北侧，景观设计根据绿地区域的几何结构来设计，设计为一个朝向城市中心逐渐延伸的方式。公园和公共空间都是城市道路的统一系统的组成部分，消弱了建筑物作为对公共区域社会性用途之障碍物的存在感。

**页码248-259**

**地点** 地拉那，阿尔巴尼亚
**项目** 商业
**客户** 地拉那市 – ManeTCI
**规划** 2009年设计竞赛，第2名

## RESIDENZA DEL FORTE CARLO FELICE
**La Maddalena, Italy, 2008-2009**

The hotel complex built on the island of La Maddalena in anticipation of the G8 summit in July 2009, includes an extensive environmental rehabilitation project that affects the transformation of the military ex-hospital building, the redesign of a stretch of coast and rear area ceded by the Italian Navy to the Region of Sardinia. The architectural project takes its cue from a previous preliminary design realized in a technical structure called "Struttura di Missione", a direct expression of the Chair of the Council of Ministers. The final solution is the attempt to "naturalize" the building making it "belong" to the land, leaving to the original early Twentieth Century structure, the role of the palace, of the actual building, while the largest part of the project confronts the scale of the landscape. The building roof goes beyond the idea of the roof-garden to take on the role of hillside park. The building is totally "embedded" in the rear promontory, opportunely dug out to to embed the enormous volumetry expected to stretch over more than 300 meters to the back of the pre-existing building.
**pag 260-295**

**location** La Maddalena, Olbia Tempio, Sassari, Italy
**project** Hotel and restaurant
**client** Presidenza del Consiglio dei Ministri - Dipartimento di Protezione Civile
**structures** MTM progetti S.r.l.
**mechanical systems** Eugenio Cimino
**electric systems** Flavio De Vito
**plan** 2008
**construction** 2008-2009
**cost** € 53,000,000
**plot area** 35,000 sq.m
**built area** 17,500 sq.m
**volume** 56,000 cu.m
**contractor** Gia.Fi. Costruzioni S.p.a.

## PADUA MASTERPLAN
**Padua, Italy, 2008-2009**

The area between Padua's fair center and the central station, north of the Tempio della Pace church, is an important part of the complex urban system by the historic center. It has been in pursuit of a definitive layout for over twenty years. The urbanization and development plan for such an important, complicated area, a space set between some of the city's most important infrastructures, is part of a strategy of major transformation that was started in the 1980s and has yet to be completed. The new masterplan that the city administration has approved chooses a new city model that works towards the density of the historic center and the city's public spaces, eliminating surface parking areas to make the areas around buildings exclusively for pedestrian use. The design approach is based on compact volumes following the traditional style of city blocks. In addition to containing a protected, private internal space, the design opens up perspective views of sites of considerable quality, such as the Tempio della Pace church and the courthouse. The vehicle ramps to the underground parking lots are located by high traffic thoroughfares and are protected by a one-way service road freeing the entire area from car traffic and surface parking lots. This means that between the buildings, instead of streets, there are extra wide sidewalks, covered by paved surfaces rather than asphalt.
**pag 296-311**

**location** Padua, Italy
**project** Master plan
**client** I.F.I.P.
**plan** 2008-2009
**plot area** 61,405 sq.m
**built area** 48,506 sq.m
**volume** 156,400 cu.m
**contractor** I.F.I.P.

## MERAVIGLIOSA ISLAND
**Dubai, United Arab Emirates, 2009**

On the one side, there is a slightly sloped beach, and on the other, a layered, organic sea shelf that simulates the rugged quality of a crag. The design envisions a grouping of 15 single-family homes on the two long sides of the island, in addition to a small building for facilities. There are three different villa models: Beach Villa (along the beach); Ocean Villa (on the artificial sea shelf); and Water Villa (islands on piles in the middle of the sea). The housing model's form and structure are a take on the idea of the tent and the desert with its dunes and its wind-shaped roughness. The villas are configured like large roofs conceived continuously with the façades, which merge with the local landscape. The materials were inspired by desert colors: terracotta in a variety of colors and colored cement; the terracotta is shaped using traditional building technologies like the Catalan vaulting technique, and the colored cement is made into curved shells shaped by metal substructures. Each villa differs in its configuration and its design, following a plan far from the speculative approach that has defined recent divisions in Dubai. Yet, each house tries to come into harmony with the adjacent building with an approach based on the defining merging and homogeneity of a desert landscape.
**pag 312-327**

**location** The World, Dubai, United Arab Emirates
**project** Residential, hospitality
**client** Rajaa Trading & Investiments
**plan** 2009
**construction** In progress
**plot area** 40,642 sq.m
**built area** 16,782 sq.m

## CARLO FELICE 堡酒店
La Maddalena，意大利，2008年-2009 年

这座酒店综合建筑物建于LaMaddalena岛上，预计用于2009年7月的G8峰会，包括了一项广泛的环境整治项目，影响到了前军事医院建筑的改造、海岸线地带以及意大利海军归还给Sardinia区的后围区域的重新设计。这个建筑项目的设计灵感来源于之前在一所称为"Struttura di Missione"上实现的技术结构的初步设计，是一种对内阁主席设想的直白表述。最终解决方案试图让这座建筑物"自然化"，使其"归属"于这座岛屿，留给原来的20世纪早期构筑物、实际建筑物的宫殿角色，同时让项目最大组成部分正面朝向整个景观区。这座建筑的屋顶部分超越了屋顶花园的思路，承担起山坡花园的角色。建筑物完全"嵌入"到后面的山脚下，即：这个预计超过300米的巨大体量的建筑物，挨着现存建筑物的背后，完全地嵌入山脚下。

页号260-295

地点　La Maddalena，Olbia Tempio，萨萨里，意大利
项目　酒店和餐厅
客户　Presidenza del Consiglio dei Ministri - Dipartimento di Protezione Civile
结构　MTM progetti S.r.l.
机械系统　Eugenio Cimino
电气系统　Flavio De Vito
规划　2008年
建设　2008年-2009年
成本　53,000,000欧元
占地面积　35,000平方米
建筑面积　17,500平方米
体量　56,000立方米
承包商　Gia.Fi. Costruzioni S.p.a.

## 帕多瓦总体规划
帕多瓦，意大利，2008年-2009年

这块帕多瓦集市中心与中央火车站之间的区域位于Tempio della Pace教堂以北，它是毗邻历史中心的复杂城市系统当中重要的组成部分。为了寻求对这块地区的权威性布局方案，当地已经花了超过20年的时间。对于如此重要又复杂的区域，一块身处这座城市最重要基础设施之间的空间，是起始于1980年代查，尚未完成的重大城市改造策略的一部分。这份已经被城市行政当局批准的总体规划选择了一种新型的城市模式，努力处理好历史性中心和城市公共空间密集布置的主题，取消地表公园区域，让围绕这些建筑物的区域仅供行人使用。此设计方法以遵循城市街区传统风格的紧凑体量为基础。除了包含了保护良好的私家内部空间之外，设计方案提供了品质相当高的建筑观察视角，例如对Tempio della Pace教堂和法院大楼。通至地下停车场的车辆斜坡道路挨着车流量较大的道路，并配有单向服务性道路，让整个区域内免于出现汽车交通流并取消了地面停车场。这意味着在这些建筑物之间，不设置街道，而是设计了超宽的人行道，人行道上采用了铺砌地面而不是沥青地面。

页号296-311

地点　帕多瓦，意大利
项目　总体规划
客户　I.F.I.P.
规划　2008年-2009年
占地面积　61,405平方米
建筑面积　48,506平方米
体量　156,400立方米
承包商　I.F.I.P.

## MERAVIGLIOSA岛
杜拜，阿拉伯联合酋长国，2009年

这个再生岛屿的线性规划，分割出两种不同的滨水区。在一侧，是一种略微倾斜的海岸斜坡，而另一侧是一种分层式的有机大陆架，模拟着悬崖峭壁的崎岖不平特质。这项设计设想在这个岛两个长边形成15个独户房屋群，并且设计了一个小型建筑物供设备使用。这里有3种不同的别墅模式：海边别墅（沿海岸布置）；海洋别墅（在人造大陆架上）；以及水中别墅（建在海中心桩架上的小岛）。房屋模式的形式和结构的创意来源于帐蓬以及沙漠沙丘以及被风塑造出来的粗糙感。这些别墅设计就像一个个大屋顶的样子，人们可以感觉到它们与建筑外表皮是连续的，而建筑表皮绵延起伏又与当地景观融为一体。材料灵感来源于沙漠的色彩：各种色彩的赤陶和彩色水泥；赤陶采用传统建筑技术塑成形，诸如加泰罗尼亚拱顶建造技巧，而彩色水泥制作成弯曲的壳体并由金属支撑结构塑形。每一座别墅的配置及设计均各不相同，所遵循的规划远离了那些在杜拜存在着明显短期的投机性方法。而且，每座房屋都试图达成与邻近建筑的和谐一致，与沙漠景观完美融合。

页号312-327

地点　世界岛，杜拜，阿拉伯联合酋长国
项目　住宅，酒店
客户　Rajaa Trading & Investiments
规划　2009年
建设　在建
占地面积　40,642平方米
建筑面积　16,782平方米

## LILING WORLD CERAMIC ART CITY
**Liling, China, 2010**

The concept for the Liling design came out of our client's wish to site certain other important functions, such as a museum and a hotel, in an industrial are for ceramics processing. The important role played by this manufacturing activity for Liling County contributes to make the ceramic product the chief formal inspiration of the compositive volumes of the design. The various functions are placed within volumes with a circular plan, or inside forms that can be geometrically circumscribed. At any rate, the number of volumes is greater than the actual functional needs (the main functions are: gate, museum, hotel, and master building). As such, by multiplying them, they can create a freely organized system that can delineate the open space and establish a relationship of vicinity that we call the "space between." Following this principle, the form and examples of possible volumes are not very important in and of themselves. Of essential importance, on the other hand, is a close look at the space of the "vases" whose vase shape delineates sinuous contours with no sharp edges, always concave and convex. The "space between," as shown in the initial concept, takes substance in a line of vases that follow a commutative rule, shifting without changing the final result, which is always ensured by the juxtaposition of the vases' generative contours. According to the system suggested in the design, density becomes a value, a resource that allows for a dense relationship, a vicinity, a use of the space on the ground like that of the historic city.
**pag 328-339**

**location** Liling, China
**client** 建设单位
**project** LiLing World Ceramic Art City
**plan** 2010
**construction** in progress
**plot area** 110000 mq
**built area** 50000 mq

## LVBO CORE CLUSTER AREA
**Zhengzhou, China, 2011**

The master plan of the New Lubo District in Zhengzhou is based on the guidelines provided by the town plan of the Isozaki firm; it features a surface area of 24 square kilometres and is conceived to accommodate about 370,000 inhabitants and a quite important financial-administrative centre, the core, with an area of about 4.5 square kilometres. The latter, characterized by an irregular European-style urban web and by the two principal landmarks of the project, a museum and a library, is surrounded by two concentric rings; the inner consists of water and green islands while the outer is residential. North of this area, separated from the rest of the master plan by a large east-west thoroughfare axis, the industrial area and technological centre of the city is located. The guiding principle of the whole project is environmental sustainability, along with a rediscovery of a public space on a human scale, characterized by tall sculptural landmarks combined with water courses and green areas. The green space is integrated with an extensive network of water courses and a residential system inspired by the traditional Chinese city and its ancient hutongs. While the traffic is optimized by an efficient network of parking areas and two main fast-traffic axes, it is essentially based on public transport and on slow movement, by foot and by bike, in order to make people use the streets again and create new places where members of the community may meet and socialize.
**pag 340-351**

**location** Zhengzhou, China
**client** Administrative Committee of Zhengzhou Zhongmu Industrial District
**anno** 2011
**plot area** 22445163 sq.m
**built area** 16564269 sq.m
**contractor** Administrative Committee of Zhengzhou Zhongmu Industrial District

## STOP LINE
**Curno, Italy, 1993-1995**

Located along the Briantea high traffic state road in a typical urban outskirt with regular divisions and industrial structures, the building to be converted was an "anonymous cement box bought to become a center for entertainment, leisure and gathering". The design identifies with the place and the "box-like" quality of its buildings, while rejecting the inexpressiveness that usually goes with such buildings. The façade's dual identity is the first impression it makes: in the daytime, the image it presents is of a slightly perforated Corten steel façade, punctuated by the openings of the escape doors and the vertical slash of the entrance, topped by a projecting cantilever roof; at night, the rust wall dematerializes and turns into a dynamic vision of light points, multiplied in the reflection of the water in front of it. The plan highlights the existing perimeter and structure and centralizes the arena like an open square surrounded by public establishments. Its activities cover four levels, including an underground level and a roof/terrace that is open to vehicles. The building has been subject to a new conversion project that returns it to use for a more anonymous business activity, linked once again to work.
**pag 354-367**

**location** Curno, Bergamo, Italy
**project** Recreational complex
**client** Golf Parco dei Colli
**structures** Studio Myallonnier
**systems** Studio Armondi
**plan** 1993
**construction** 1995
**cost** € 11,000,000
**built area** 10,000 sq.m
**volume** 60,000 cu.m
**contractor** Falgari Mario & C. S.n.c. – F.lli Bergamini

# 醴陵世界陶瓷艺术城

醴陵，中国，2010年

醴陵设计的概念源于客户希望在陶瓷加工工业区内设置某些其它重要功能，例如博物馆和酒店。而通过制造业活动为醴陵县所发挥出来的重要作用则有助于使得陶瓷产品成为本设计组成部分中的灵感源泉之所在。

各种功能设施通过一个圆形平面被囊括进了建筑物之中，或被安置进了几何式划定范围的建筑形式之中。但无论如何，建筑物数量大于实际的功能需求（主要功能设施有：大门、博物馆、酒店和主建筑）。因此，通过这些功能设施的叠加，将可以创建一套自由式的有组织系统，从而能够划定开放空间并构建我们称之为"间距"的邻近关系。

遵循这一原则，可能性建筑物的形式和范例就其自身而言并不是非常重要。从另一方面来说，其中的根本重要性在于寻找"花瓶"处的一个紧凑视觉点，而该花瓶形状则表现为弯曲的轮廓但无锋利的边缘，且总显凹凸不平。

该类"间距"，如最初概念图所示，实现了物体与花瓶线条的走向一致并遵循交换及位移的原则，但以不改变最终结果为前提，且该目的始终通过花瓶生成性轮廓的并置来加以保障。

根据设计中推荐的系统内容，密度成为了一种评估值，而资源则考虑到了密集关系、邻近地区、以及如同对待历史名城态度般的对空间的利用。

pag 328-339

**地点**　醴陵，中国
**客户**　建设单位
**项目**　醴陵世界陶瓷艺术城
**规划**　2010年
**建设**　建设中
**占地面积**　110000平方米
**建筑面积**　50000平方米

# 绿博组团核心区

郑州，中国，2011年

T坐落于郑州市新陆博区的本项目的总体规划是基于矶崎公司城镇规划中的指导方针的，而这则意味着公开追求竞争的始发点；该项目地表面积为24平方千米，设计容纳能力大约为37万居民，涵盖一个相当重要的金融行政中心，且该核心区域的面积大约为4.5平方千米。而后者，以不规则的欧式风格城市布局以及本项目的两个主要地标，即：博物馆和图书馆，为其特征，被两个同心环路所围绕；内环以水道和绿洲作为构造而外部则为居民区。而本地域的北面，经由一个大的东西横贯通道轴将其分割在总体规划之外，是为城市工业区和技术中心所在地。

整个项目的指导原则是环境的可持续性，以及对人类活动公共空间的重新认知，而本项目的特征为高大雕塑地标汇聚水道和绿地。而绿地区域则汇集了纵深的水道网络以及采撷传统中国城市及其古老胡同文化精髓的住宅系统。

交通则通过停车区的高效网络化以及两个主要快速交通轴来加以优化，它从本质上而言是基于公共交通和慢活风潮，可通过步行和骑自行车的方式，以便让人们再次踏上街道，并创建新的社区居民可以聚集并参与社交的处所。

pag 340-351

**地点**　郑州，中国
**客户**　郑州中牟产业园区管委会
**规划**　2011年
**占地面积**　22445163 平方米
**建筑面积**　16564269 平方米
**承包商**　郑州中牟产业园区管委会

# 停止线

Curno，意大利，1993年-1995年

所要改建的这座建筑物位于Briantea的大交通量国道之旁，位于典型的拥有规则分区和工业厂房的城市郊区，改建宗旨是"将无足轻重的水泥盒子变成一个娱乐、休闲和聚会中心"。设计方案突出表现这个场所以及其建筑物"盒子似"的特质，同时排除了这类建筑物所具有的表现能力呆板的通病。建筑立面给人的第一印象是其双重身份性：在白天，所呈现的形象是略有一些穿孔的考登钢外表皮，并被火灾逃生门的开口以及垂直的入口所穿透，顶部是一个突出的悬臂式屋顶；在夜晚，生锈的墙壁呈现去物质化的效果，转化成光点斑驳的动态景象，并在墙壁前面的水池倒映中倍增显现。设计方案强调了现有的建筑周边部分和结构，舞池设在中心区，如同被公共设施所围绕的开放式广场。此建筑分为4层，包括了地下层和开放给车辆的屋顶和梯阶。此项目的改造工程，再一次悄然声息的将商务与工作紧密的联系起来。

页号354-367

**地点**　Curno，贝加莫，意大利
**项目**　娱乐综合建筑物
**客户**　*Golf Parco dei Colli*
**结构**　*Studio Myallonnier*
**系统**　*Studio Armondi*
**规划**　1993年
**建设**　1995年
**成本**　11,000,000欧元
**建筑面积**　10,000平方米
**体量**　60,000立方米
**承包商**　*Falgari Mario & C. S.n.c. – F.lli Bergamini*

## POOL IN MONTE BUE
**Cene, Italy, 1995-1996**

The project is located along the side of the mountain and is the extension of a pre-existing villa park, for the realization of an open-air swimming pool. The natural shape of the slope suggested the idea of an artificial soil containment buttress in order to find a new horizontal level on which to locate the pool. The pool is designed as a system of two reservoirs: one rectangular for cold water, the other one smaller, with heated pressure water jets, which follows the configuration of the rampart. The image of the buttress is emphasized by a walkway which follows the perimeter of the stone terracing allowing you to lie on the grass without seeing the protective wall which is consequently at the same height as the garden. The attachment to the hill is instead designed from the rocky wall and from the rocks obtained from the excavation necessary to obtain a level coplanar with the tank. Such a ground subtraction operation has shown, below the cultivation level, a stone fault hence reused to create a naturalized perimeter around the swimming pool. The construction blocks, worked on site, have been squared in relation to the veining and positioned according to the project to define the terraced wall which conceals both the machines and containment walls of the dressing room and a small kitchen.
**pag 368-375**

**location** Cene, Bergamo, Italy
**project** Pool
**client** Private
**plan** 1995
**construction** 1996
**cost** € 300,000
**contractor** Madaschi

## SQUARE AND PUBLIC FACILITIES
**Merate, Italy, 1996-2011**

The site is set between the established city and new developments near Merate's historic center. The instructions for the project for this area, which covers more than a square hectare, were set in the city's general urban plan and by the public who had opposed an earlier proposal to turn the area into a service industry district. The project includes an underground parking lot and a landscaped green area compatible with the established urban fabric and its emerging features. The proposal keeps to the idea of a park and a square, maintaining the built city's spirit without imposing its own volumes; a system of rampart/terraces navigates the changes in levels with stairways and ramps, organizing the square's area in panoramic levels. Making use of the natural shifts in levels and the spaces freed from the containment walls for the new roofs covered with a simple grass layer, the project creates an outdoor theatre, spaces for cultural organizations and a café, plus an underground parking lot with over 130 spaces. The greenery of the lawn and the trees combines with vertical and sloping surfaces that are covered in hand-made brick that is narrow and elongated, that gives uniformity and relates the new square to the building tradition of the city's central tower.
**pag 376-389**

**location** Merate, Lecco, Italy
**project** Public square and public facilities
**client** Municipality of Merate
**structures** Studio Myallonnier
**systems** Studio Armondi
**plan** 1996
**construction** 1999-2011
**cost** € 3,200,000
**plot area** 11,260 sq.m
**built area** 5,280 sq.m
**contractor** CIAS Group - SGC Italia S.p.A. - Giacomo Zenga Costruzioni S.a.s

## CURNO LIBRARY AND AUDITORIUM
**Curno, Italy, 1996-2009**

The site for the building of a new public library and a small auditorium with 250 seats has been found inside an existing school campus. The location has suggested the idea of a project centred on the continuity of the surrounding public area. The building, conceived as a kind of open book, is therefore characterized by a sloping roof which is terraced to form stands, and which may be used for open-air events, to extend the public square in front to the roof of the building. The plan distinguishes the activities which pivot on the rectangular hall of the auditorium from the areas of the library, whose perimeter consists of a longitudinal outline characterized by the jagged geometry of the external facade. The line of demarcation and communication between the two areas takes the form of a new urban itinerary, a triple-height void paced horizontally by a system of split levels which in their turn serve as communication paths, and vertically by a succession of uprights which structure the entire wall as container case of books. The materials, which have been reduced to the essentiality of an untreated concrete mixed with colour, have made it possible to mould the vertical surfaces as the pages of a conceptual book, engraved here and there with letters.
**pag 390-411**

**location** Curno, Bergamo, Italy
**project** Library and auditorium
**client** Municipality of Curno
**structures** Studio Myallonnier
**systems** Studio Armondi
**plan** 1996
**construction** 1999-2009
**cost** € 2,000,000
**built area** 1,960 sq.m
**volume** 8,200 cu.m
**contractor** Viola Costruzioni

## MONTE BUE游泳池
**Cene，意大利，1995年-1996年**

这座露天游泳池位于山坡之旁，是一座别墅花园的扩建部分。这个山坡的天然地形提示了采用人造土壤保固扶壁来寻求新的水平面来设计此游泳池的思路。这座游泳池设计为一个由两个蓄水池构成的系统：一个长方形蓄水池用于冷水，另一个小一些的蓄水池配备了加热加压水流喷头，这些喷头随池壁形态而布设。扶壁的外形由一条步道来突出强调，这条步道沿石砌阶梯的周边布置，人躺在草地上时是看不到保护墙的，因为保护墙的高度与花园相同。与山丘的依附关系在设计上反而来源于岩石墙壁并来自于开挖时所获得的岩石，而开挖是为了获得与水池共平面标高而必需进行的工程。这样便利用绿植层以下的石料断层为泳池提供了一个天然的镶边。在现场制作的建筑体块根据纹理修整成了正方形，并按照项目要求进行定位，形成了梯级墙壁，将设备和更衣室，小厨房的挡土墙都隐蔽了起来。
**页号368-375**

**地点**　　Cene，贝加莫，意大利
**项目**　　游泳池
**客户**　　私人客户
**规划**　　1995年
**建设**　　1996年
**成本**　　300,000欧元
**承包商**　Madaschi

## 广场和公共设施
**Merate，意大利，1996年-2011年**

项目位于已经建成的城市和新开发区之间，位置靠近Merate的历史中心。这块占地超过1公顷的区域的规划方针按照这座城市的总体城市规划来设定，而公众反对之前的将这块区域改造成服务业街区的方案。项目包括了一个地下停车场和景观化的绿地，此绿地与确立已久的城市基本结构和新出现的特色相兼容。方案保留了这个城市的精神主旨——公园和广场的设计创意，而不是强迫插入建筑体块；一个由台阶和坡道控制高度变化的高台/露台系统，将广场全景在不同的高度上呈现。通过自然高差和从挡土墙中释放出的空间，形成采用简单草地层的新屋顶，构造了一个室外剧场、文化活动空间以及一座咖啡馆，加上拥有超过130个车位的地下停车场。草坪和树木的绿化效果结合了手工制造的形状狭长的砖块覆盖的垂直和斜坡表面，这种均一材料的使用，让新广场延续了市中心高塔的传统施工方式。
**页号376-389**

**地点**　　Merate，莱科，意大利
**项目**　　公共广场和公共设施
**客户**　　Merate市政府
**结构**　　Studio Myallonnier
**系统**　　Studio Armondi
**规划**　　1996年
**建设**　在建 1999年-2011年
**成本**　　3,200,000欧元
**占地面积**　11,260平方米
**建筑面积**　5,280平方米
**承包商**　CIAS Group - SGC Italia S.p.A. - Giacomo Zenga Costruzioni S.a.s

## CURNO图书馆和礼堂
**Curno，意大利，1996年-2009年**

用于建设一座新公共图书馆和250座小型礼堂的建址位于现有大学校园内。这个位置促使我们形成了让项目位于周围连续性公共区域中心区的设计思路。这座建筑物以类似于一种打开的书籍的形象作为构思，阶梯式的屋顶，作为露天广场使用。建筑物前方的公共广场延伸至阶梯式屋顶，并且尽量减少对土地的占用。平面为三角形的阶梯下，长方形的礼堂从平面图上清晰辨认。图书馆区周边由纵向轮廓线构成，其特色是外部正面锯齿状的几何图案。这两个区域之间的界定和联系采用了新城市路线的形式，三层通高的中庭，设有水平向的，在不同高度的走廊，服务于交流，连接，或者垂直交通。这些结构的外部是书籍形象外壳的完整墙壁。材料精简到只用彩色清水混凝土，并随机雕刻着一些英文字母，使得我们可以让垂直表面模制成类似于一本概念性书籍的书面。
**页号390-411**

**地点**　　Curno，贝加莫，意大利
**项目**　　图书馆和礼堂
**客户**　　Curno市政府
**结构**　　Studio Myallonnier
**系统**　　Studio Armondi
**规划**　　1996年
**建设**　　1999年-2009年
**成本**　　2,000,000欧元
**建筑面积**　1,960平方米
**体量**　　8,200立方米
**承包商**　Viola Costruzioni

## LEFFE HOUSE
**Leffe, Italy, 1997-1999**

The building, in the old town of Leffe in the Seriana Valley, is the result of the demolition and reconstruction of an existing house. The plot, defined on the sides by two adjacent buildings with different façades narrows down from about 10 metres on the side facing the street to only 5 metres on the side overlooking the valley. The condition of the site and the distances which had to be respected to the complex arrangement of the facing buildings have suggested the idea of a jagged and folded main façade, dominated by the tower-shaped outline of the staircase. Placed in close contact with the facades of the other buildings the front forms a solid curtain with the exception of strips of light which enter the interior from a sequence of rifts cut into the masonry. The rear facade is characterized by the glazing of the entire space enclosed by the neighbouring buildings. This has called for the design of a system of screens, formed of large metal doors which fold together to offer an unobstructed view of the valley, and to protect the entire surface of the façade. This mobile partition echoes the theme of the long and narrow openings characterizing the opposite facade, at the same time reminding of the typical sunscreen of traditional farmhouses and barns.
**pag 412-427**

**location** Leffe, Bergamo, Italy
**project** One-family house
**client** Private
**structures** Gianfranco Calderoni
**plan** 1997
**construction** 1999
**cost** € 600,000
**plot area** 90 sq.m
**built area** 280 sq.m
**volume** 800 cu.m
**contractor** Madaschi

## RESIDENTIAL AND COMMERCIAL COMPLEX
**Tavarnuzze, Italy, 2000**

The plot is located between the Greve river, the Cassia highway and the centre of Tavarnuzze. The project – dwellings above shops on the ground floor – meets the need to create public spaces and itineraries to revitalize the urban fronts along the highway. The volumes are the result of the demolition of old sheds used to store fuel. The main volume follows the curve of the river where the smaller extremity takes on the proportions and characters of the ancient building in front. The large stairs on the main facade lead to the level where the entrances to the dwellings are located, and where a longitudinal public square hosts new urban itineraries and rest areas. The first floor, which is the roof of the shopping area below, is detached from the ground, with a transversal projection of 7 metres. The hard base in stone appears dematerialized thanks to the facing in acid-treated copper which creates a uniform volume including the prominent pitched roof: the metal facing evokes the chromatic shades of the vegetation, while the windows and other openings are arranged on the façades, as if the construction had been built by subsequent layers.
**pag 428-439**

**location** Tavarnuzze, Florence, Italy
**project** Dwellings and shops
**client** Bruno Cecchi S.p.A.
**structures** Vega Ingegneria
**mechanical systems** Studio Mancini
**electrical systems** Xenia s.r.l.
Alessandro Lepri
**plan** 2000
**construction** In progress
**cost** € 8,000,000
**plot area** 6,600 sq.m
**built area** 5,256 sq.m
**volume** 17,500 cu.m

## MERATE TOWN HALL
**Merate, Italy, 2001**

The project centres on the adaptation of the old Town Hall of the city, built in the late Nineteenth century, by the addition of a long and narrow wing on the rear, placed orthogonally to the original building. In order to restore the urban status of the aggregate to a level adequate the project provides an adaptation of the distribution of the existing building, to which new public spaces have been added. The complex now vaunts town hall, auditorium, offices and premises for the traffic police. The project features the addition, on the rear of the building, of three narrow and elongated wings. The central axis formed by the entrance leads directly from the square in front of the old building to a T-shaped foyer which communicates with the new functions and activities housed by the aggregate. A large triple-height space which evokes the "public galleries" of the Nineteenth century connects the existing building and the new linear volumes placed perpendicularly to it. The latter appear as semi-transparent shells that conceal and reveal the interiors through the decorative web of the external sunscreens formed by a system of "grates" in cast aluminium, fixed to the load-bearing structure, which protect all glazed parts.
**pag 440-457**

**location** Merate, Lecco, Italy
**project** Town hall and auditorium
**client** Municipality of Merate
**structures** Studio Myallonnier
**systems** Studio Armondi
**plan** 2001-2005
**construction** Under construction
**cost** € 3,615,200
**built area** 4,725 sq.m
**volume** 21,700 cu.m
**contractor** Giacomo Zenga Costruzioni s.a.s - Impianti Tecnologici Sbrescia s.n.c.

## LEFFE 别墅
**Leffe，意大利，1997年-1999年**

这座建筑物位于Seriana山谷的Leffet老镇，是现有房屋拆除和重建的成果。这个地块的两侧界限由两座拥有不同建筑外立面的邻近建筑物来界定，从面向街道一侧的10米左右宽度收窄至俯瞰山谷侧的仅5米宽度。建址的条件和间距必须考虑到对面建筑的复杂布局，这些促使我们形成了一种锯齿状和折叠式的主建筑正面的设计思路，这个正面由具有塔形轮廓线的楼梯间所主导。在布置上与其它建筑物正面密切接触的情况下，正面形成了一种厚实的帷幕，从砖砌结构上切出的狭缝序列让光以条带形式进入到建筑内部。背立面整体采用玻璃窗形式，在邻近建筑物的自然围合空间中尤显特色。这就要求设计一套屏风系统，由大型金属遮板构成，折叠起来后就能获得的毫无阻碍的山谷观景，并且能保持立面的整体性。这个活动的分隔结构与正立面所特有的狭长主题相呼应，同时可令人联想到传统农舍和谷仓的典型遮阳蓬。

**页号412-427**

**地点** Leffe，贝加莫，意大利
**项目** 独户房屋
**客户** 私人客户
**结构** Gianfranco Calderoni
**规划** 1997年
**建设** 1999年
**成本** 600,000欧元
**占地面积** 90平方米
**建筑面积** 280平方米
**体量** 800立方米
**承包商** Madaschi

## 商住综合楼
**Tavarnuzze，意大利，2000年**

此地点位于Greve河、Cassia高速公路与Tavarnuzze中心之间。此项目为底商住宅楼，它满足了能够让沿高速公路的城市正面重新恢复活力的公共空间和路线的需要。这个建筑体量来源于用于贮存燃料的旧棚式建筑。主要体量沿河流的曲线设置，较小的末端继承了前面古代建筑的比例和特点。主立面的大楼梯导向住宅所在楼层的入口，而且这里纵向的公共广场会成为新的城市观光区和休憩区。第二层，也就是底下商店区的屋顶，与地面是分离的，并存外挑7米。石制的硬地基显示出物化的效果，这是由于表面采用了经酸蚀处理的铜材，形成了一种均匀分布的体量效果，包括醒目的斜屋顶：这种金属饰面会给人植被多重阴影色彩的感受，而窗户和其它开口也安排在建筑立面上，就像这个构筑物是由后来的构造叠加而成。

**页号428-439**

**地点** Tavarnuzze，佛罗伦萨，意大利
**项目** 住宅和商店
**客户** Bruno Cecchi S.p.A.
**结构** Vega Ingegneria
**机械系统** Studio Mancini
**电气系统** Xenia s.r.l. Alessandro Lepri
**规划** 2000年
**建设** 在建
**成本** 8,000,000欧元
**占地面积** 6,600平方米
**建筑面积** 5,256平方米
**体量** 17,500立方米

## MERATE市政厅
**Merate，意大利，2001年**

项目的中心是对Merate市旧市政厅的改造，旧市政厅建于19世纪后半叶，改造时在后部添建了狭长的翼部，与原有建筑成正交。为了将这个建筑集合体的城市状态恢复到一个足够的水平，项目提供了对现有房屋分配的改造，并加建了新的公共空间。这座建筑群如今拥有了市政厅、礼堂、办公室和供交通警察使用的房屋。项目的特色体现在建筑物后面添加的3个狭窄而拉长的翼部。由入口构成的中轴线直接从至旧建筑物前方的广场导向至一个T形的门厅，这个门厅让这个建筑集合体所容纳的新职能和活动相通起来。一个大型的3层通高的空间会唤起人们对于19世纪"公共长廊"的记忆，它将现有建筑物与新建的正交线性体量连接起来。新建体量以半透明的壳式建筑呈现，通过外部遮阳棚的装饰性网格板将室内空间若隐若现的呈现出来，这个遮阳棚由一个铸铝"格栅"系统构成，固定在承重结构上，并保护着所有的玻璃部件。

**页号440-457**

**地点** Merate，莱科，意大利
**项目** 市政厅和礼堂
**客户** Merate市政府
**结构** Studio Myallonnier
**系统** Studio Armondi
**规划** 2001年-2005年
**建设** 在建
**成本** 3,615,200欧元
**建筑面积** 4,725平方米
**体量** 21,700立方米
**承包商** Giacomo Zenga Costruzioni s.a.s - Impianti Tecnologici Sbrescia s.n.c.

## PORTA SUSA STATION
**Turin, Italy, 2001**

The theme of the restricted selection international competition involves interpreting Porta Susa station as a boundary place, a true city gate, and hub for transport and services. The idea is thus to give the passenger building not only specific functions related to the waiting areas and transit but also for cultural and commercial exchange. The project takes on the geometric configuration of its narrow and elongated forms directly from the area's dimensions, while the memory of the original tracks remains as a trace in the design of the facades and in the suppleness of the development plan. Parallelism and interweaving of lines resulting in incisions, cuts, ramps, volumetric juxtapositions interpenetrating between ground and building. Gaps and disconnections maintain a visual and physical continuity internally, due to a studied series of structural elements and to double and triple height balconies that allow distribution of natural light. The perception of the space, along the axis of the railway and the flow of the paths makes the new station more an urban journey than a building. The access points invite to the length of a sort of gallery integrated with the surrounding fabric while the symbolic element of the tower, split into two monoliths, interact visually with the features of the historical city.
**pag 458-467**

**with Studio Pellegrini, Efisio Pitzalis and Studio Gambogi**
**location** Turin, Italy
**project** Infrastructure
**client** Ferrovie dello Stato, transport and action services company, infrastructure division
**structures** Studio Gambogi
**plan** 2001 competition

## POOL IN GAZZANIGA
**Gazzaniga, Italy, 2001-2002**

This very small project deals with the transformation of one of the many small villas that characterize the north Italy valley floors. The significant aspect of the project concerns the commonality of the starting structure, one of the classic single-family houses, arranged on two or three levels, of which one half-basement dedicated to the games room. Externally the half-basement is hidden by a small manmade hill that surrounds the house, a kind of artificial landscape. The project's objective is to destroy the fake little hill, dig it out, give light back to the games room, as the true heart of the house, building a new landscape and a new panorama that reflects on the pool, designed as the new internal patio around which the house winds. This kind of secret garden is concieved of as a room where the floor is the pool that reflects the blue of the sky. The new patio is surrounded by a portico featuring steel columns that in its turn covered by ground that re-creates the garden above the ground, supports the sliding window tracks and the terracotta brise-soleil panels. The spatial continuity between the exterior and interior is further guaranteed by a travertine-paved surface of that covers the entire half-basement level – now a real ground floor – including also the uncovered portion of the walkway and of the solarium. A set of green steps connect the garden level with the lower pool level, while local stone walls support and contain the natural unevenness of the ground.
**pag 468-475**

**location** Gazzaniga, Bergamo, Italy
**project** Residential
**client** Private
**plan** 2001
**construction** 2002
**cost** € 800,000

## EX AREA FIAT RESIDENTIAL COMPLEX
**Florence, Italy, 2001**

The former Fiat area in Novoli, northwest quadrant of Florence, has been since the eighties subject of a complex recovery program on which many international architects have worked according to Leon Krier's master plan which set the building criteria and rules organized around the redesign of the traditional block.
Within this plan, later developed by Gabetti and Isola, Archea, on behalf of the Novoli property company that manages the development program, has realized over the years, three different design solutions. The first regards a lot situated in the immediate vicinity of the Palace of Justice and dedicated to offices, the second and third options are on a different lot, overlooking the new San Donato park, meant for residences. All three projects adhere to the common rules that provide a ground floor for shops, pitched roof, mainly vertical rectangular windows, and the use of traditional ceramic or stone covering materials for exterior surfaces. In the first, the interpretation of the local architectural character is translated into a single body carved out through the special excavation of the volume that transforms the office building into a "virtual" placement of blocks characterized by a consistent urban dimension. This strategy is designed to identify, in the one building block, several sub-buildings, covered by a pitched awning, without indulging in the vernacular medieval-looking plan suggested by Krier. The single material volume proposes an unusual planimetric reading of the whole offered by the high view of the building possible from the windows of the above Palace of Justice. The roofing is consequently interpreted as an elevation, treated with the same cladding (red Persian travertine in an early version, then

## PORTA SUSA车站

都灵，意大利，2001年

这项严谨的国际设计竞赛主题是诠释Porta Susa火车站作为边界场所、一个真正的城市大门以及运输和服务中心的意义。因此设计创意就是不仅赋予这座旅客车站与候车区和转乘相关的具体功能，还要给予它文化和商业交流的职能。项目场地的尺寸决定了建筑物狭窄、细长的造型形式，同时在建筑正面的设计中以及开发规划当中保留了对原有铁轨记忆的痕迹。线条的平行和交织融入于在地面与建筑物之间的切口、开口、斜坡、体量并置穿插。缝隙和中断处在内部保持了视觉和实体的连续性，这是因为基于对结构构件系列的研究，并设置了双倍和三倍高度的阳台，让自然光进入并均匀分布。对这个沿铁路中轴线和路径流向布置的空间的视觉感受使得这座新车站更像是一次城市旅程而不是一座建筑物而已。入口点导向了一段类似于画廊，与周围环境相融的结构，与此同时，高塔的标志性元素，与分离的两个单体，在视觉上与这座历史性城市的特色实现了互动。

页号458-467

**与Studio Pellegrini, Efisio Pitzalis和Studio Gambogi合作**
**地点** 都灵，意大利
**项目** 基础设施
**客户** 国家铁路局，运输和运行服务公司，基础设施分部
**结构** Studio Gambogi
**规划** 2001年设计竞赛

## GAZZANIGA游泳池

**Gazzaniga**，意大利，2001年-2002年

这是个规模比较小的项目，涉及到许多小型别墅的改造，这些别墅是意大利北部山谷谷底的特色建筑。这个项目的主要意义是基于对其起始基本结构的思考 – 经典的独户房屋之一，按两层或三层建造，还有一个半地下室专门用作娱乐室。从外部看来，这个半地下室隐藏在一座围绕着房屋的人造小山之后，也属一种人造景观。项目的目标是彻底去掉这座小型的假山，将其挖低，让照明重回作为房屋真正核心的娱乐室，建设一个倒映在游泳池的全新风景，即一个建筑环绕的内部庭院。这种秘密花园在构思上就像一个房间一样，它的地板就是游泳池，倒映着蓝天。新露台周围环绕着采用了钢柱的柱廊，这个柱廊再由重新在地表上形成花园的地面所覆盖，支撑着滑动窗户的轨道以及赤陶遮阳板。房屋外部与内部的空间的连续性通过一个铺砌凝灰石的铺地来进一步实现，这个铺地覆盖了整个半地下室楼层，如今已经是真正的底层，还包括了走道和日光浴室的未覆盖部分。一组绿色梯级将花园层与底下的游泳池层连接起来，而用本土石材墙壁支撑并融入于天然起伏的地形。

页号468-475

**地点** Gazzaniga，贝加莫，意大利
**项目** 住宅
**客户** 私人客户
**规划** 2001年
**建设** 2002年
**成本** 800,000欧元

## EX AREA
## 菲亚特住宅建筑群

佛罗伦萨，意大利，2001年

自从20世纪80年代以来，位于Novoli的前菲亚特工厂区，佛罗伦萨的西北部，一直在进行着一项复杂的修复计划，已经有许多国际建筑设计师在根据Leon Krier的总体规划开展工作，这个总体规划设定了围绕着这个传统城区重新设计所要采用的建筑标准和规范。随后由Gabetti和Isola，Archea参与设计这个项目，Novoli物业公司负责管理开发这个项目。若干年内，形成了3个不同的规划设计方案。第一个方案涉及到一个位置紧靠法院并专门用于办公楼的地块，第二个和第三个备选方案涉及一个不同的地块，此地块俯瞰着新San Donato公园并用于住宅。所有三个项目都遵守了共同的规则，这些规则要求提供开办商店用的底层，采用斜屋顶，使用主要垂直的长方形窗户，并在外表面上使用传统的瓷制或石制贴面材料。在第一个方案中，对当地建筑设计特色的解读转化成了一种通过对体量的特殊开挖手段雕刻出的单体建筑物，将这个办公楼改造成了一种"虚拟"街区布置方式，其特色体现在连续一致的城市尺度。这种策略在设计上用于在一个建筑区块内标识出若干

terracotta in a second one) of the vertical surfaces, where the windows are simple openings positioned in order to lighten the building's overall mass. The second project for the Novoli area, lies in a different lot that opens onto the park and is characterized by the invention of a public path derived from the feet of three monolithic blocks in which the starting residential block is subdivided. The project is conceived of like an act of material excavation, the placing together of three different rocky elements that flank the park: the result is a multi-faceted volume onto which cuts and ravines open. The plastic strength of the stone-covered mass, utilized to interpret, once again, the local building tradition, is relieved by alternating color plates that make up the surfaces, both from the loggia openings topped by a copper clad roof, folded like an origami in its different roof pitches.

The third proposal, which is in fact a substantial variation of the second project, located temporally at a later phase realizing the section in which some areas located near the edge of the road are finished and the completion of the park that constitutes the prerequisite for the place's effective livability. The new building does not try to superficially or formalistically reclaim a connection with the nineteenth century city, nor least of all with the poetic visions of an "outside the town" landscape, showing instead a single and real interest thematicized in the opening of the building onto the park. The project consequently develops according to a traditional "C" or open court pattern so as to express outwardly, meaning toward the adjacent street from the lot-block, a stereometric vision represented through stone elevations, serial and closed, while, towards the inner

courtyard, that "watches" the green, the building offers a more domestic building size that appears organized and variegated in a spontaneous placement of different architectural elements, like starting from the inside of traditional urban blocks. Here the living areas on different levels face toward the open landscape through large windows – framed in a continuous wood covering like in a sort of finely designed piece of furniture – and cantilevered loggias. At the summit, it recovers both the traditional type of pitched roof, as well as the modern flat roof, that engage each other at the roofline, an alternating series of visual "telescopes", also at double height, which, alternating the recesses and jutting volumes identifies a sequence of small private gardens and livable terraces.
**pag 476-489**

**location** Florence, Italy
**project** Residential complex
**client** Immobiliare Novoli S.p.A.
**plan** 2001
**construction** In progress
**cost** € 6,000,000
**buit area** 4,500 sq.m
**volume** 14,000 cu.m

# NEMBRO LIBRARY
## Nembro, Italy, 2002-2007

The project consists of the renovation of a building from the late Nineteenth century in the old centre of a small town in the province of Bergamo, which had initially been built as a primary school. The intention was to make the building available to the citizens, by renovating and expanding the original building, which was to become the new municipal library and thus a centre of culture. The C-shaped plan of the original building and the fact that more space was needed suggested the addition of a new block on the open side, to create an internal open court and turn the building into a more stately "palazzo" formed around a court. The new volume is only connected through the basement, while it maintains a studied physical and morphological distance from the existing building. The new structure takes the form of a triple-height book-case, contained in transparent shell or casket, protected by sunscreens formed of terracotta books supported by a system of steel profiles which screen and filter the sunlight. This "diaphragm", characterized by the free rotation of the books, symbolically defines the character and the significance of the entire building.
**pag 490-513**

**location** Nembro, Bergamo, Italy
**project** Library and cultural centre
**client** Municipality of Nembro
**structures** Favero&Milan Ingegneria
**temperature control systems**
Studio Tecnico Zambonin
**electrical systems** Eros Grava
**plan** 2002
**construction** 2005-2007
**cost** € 1,888,250
**built area** 1,875 sq.m
**volume** 11,200 cu.m
**contractor** Zeral S.r.l. Costruzioni edili

个子建筑，由一个倾斜天篷覆盖，而没有过度地采用Krier所建议的当地中世纪外观的规划。单一材质的体量提示出对整体的不寻常平面解读，从法院大楼的窗户可以眺望到这种效果。屋顶也随之形成为一个立面，对垂直表面采用相同的覆层进行处理（早期设计为红色的波斯凝灰岩，然后是第二次设计的赤陶），而窗户采用简单的开口，其位置设置以照亮建筑物整个体积为目的。Novoli区的第二个项目处于一块不同的地址上，其特点是由3个建筑体量延伸出公共步道系统，项目的构思是切割原材料，建筑像是依傍在公园里的三块巨石，上面有开口和割缝，并以石材贴面加强其塑性。传统建造方式再次在多色石材贴面上，在柱廊开口和房顶上的铜瓦片上得到体现。第三个建议方案事与第二个完全不同，临时性地用于实现这个区段的后期部分，这些区段内有一些位置靠近道路边缘的区域已经竣工，而与停车场的交织构成了这个场所有效适居住的前提条件。新建筑并没有尝试肤浅地或形式主义地宣称与这座19世纪城市的联系，尤其没有提供一种"城外"景观的诗意视象，反而显示了一种在这座建筑物面向公园开口内主题化的单一而真正的趣味。因此，项目按照传统的"C"型或开放庭院模式进行开发，体现对外表达的要素，意味着对来自本地块街区的邻近街道一种立体式的景象，这个景象通过即串联又封闭还同时朝向内部庭院的石材立面来体现，

这个景象"凝视"着绿地，这座建筑物提供了更为家庭化的建筑物尺度，通过不同建筑元素的设置而呈现其组织性和多样性，类似于从传统城市街区内部开始的方式。这样处于不同层的生活区就能通过大窗户面朝开阔的景观，窗户框架采用具有连续性的木质覆层，就像一种经过精细设计的家具，以及突出的▌廊。屋顶形式包括传统类型的斜屋顶以及现代的平屋顶，在屋顶轮廓线上建有"望远镜"式大玻璃窗，有时为双层通高，高低交错，以凹凸体量的交替刻画出一个包含小型私密花园和适居阳台的建筑序列。

**页号476-489**

| | |
|---|---|
| **地点** | 佛罗伦萨，意大利 |
| **项目** | 住宅建筑群 |
| **客户** | Immobiliare Novoli S.p.A. |
| **规划** | 2001年 |
| **建设** | 在建 |
| **成本** | 6,000,000欧元 |
| **建筑面积** | 4,500平方米 |
| **体量** | 14,000立方米 |

这个项目是对一座19世纪后半叶建筑的翻新，此建筑位于贝加莫省一座小镇的旧中心，最初按一所小学建造。项目意图是通过翻新和扩建原有建筑，让这座建筑物可供市民使用，并将变成新的市立图书馆，还将因此成为文化中心。原有建筑的C形平面以及需要更多空间的事实提示我们，需要在空地一侧添建新的建筑区块，设计一个露天内庭，围绕庭院的建筑物构成更为壮丽的"宫殿式建筑"。新体量只通过地下室相连接，并同时与现有建筑之间保持一个经深思熟虑的实体和形态上的间距。新建筑采用了一个三层高的活动书橱的形式，全透明壳体外部为一套由钢结构支撑的赤陶土的书型结构的百叶，用于过滤，屏蔽阳光。这堵"隔板"的特色是书形百叶可自由旋转，从符号上定义了整个建筑物的特色和意义。

**页号490-513**

| | |
|---|---|
| **地点** | Nembro，贝加莫，意大利 |
| **项目** | 图书馆和文化中心 |
| **客户** | Nembro市政府 |
| **结构** | Favero&Milan Ingegneria 温度控制系统公司 |
| **事务所** | Tecnico Zambonin |
| **电气系统** | Eros Grava |
| **规划** | 2002年 |
| **施工** | 2005年-2007年 |
| **成本** | 1,888,250欧元 |
| **建筑面积** | 1,875平方米 |
| **体量** | 11,200立方米 |
| **承包商** | Zeral S.r.l. Costruzioni edili |

# CDD-CENTER FOR DISABILITY
## Seregno, Italy, 2003

The project area is a plot behind a nursery school, alongside an area which will become a public park, located in a disorganised residential area. The functional program, aimed at disabled persons, features primary and complementary activities – classrooms and workshops – conceived as the structures which are essential to the conduction of the socio-educational activities associated with the presence of disabled persons. The project forms a special relationship with the surroundings, hypothesizing a park accessible by a wheelchair user because the trees "perforate" a sill in smooth concrete, the pavement of this kind of artificial forest. The area has thus been redesigned as the natural continuation of the park, which bends to form the architectural volume whose sill is transformed from pavement to roof. One façade is thus open while the other is closed with the exception of the cut which defines the entrance, which can be reached by a ramp accessible to vehicles. The rectangular plan develops along the corridor communicating with the various rooms, which are also connected externally by a covered path.

pag 514-523

**location** Seregno, Milan, Italy
**project** School
**client** Municipality of Seregno
**structures** Matteo Fiori and Luca Varesi
**systems** Studio Armondi - StudioTi
**plan** 2003
**construction** Under construction
**cost** € 2,700,000
**area of plot** 10,000 sq.m
**built area** 1,300 sq.m
**volume** 4,940 cu.m
**contractor** T.i.e.c.i. S.r.l.

# THE CORD
## Venice, Italy, 2003

"The Cord" is an installation/entrance for the 50th Venice Biennale art exhibition. Somewhere between sculpture and architecture, the work's official purpose was to mark the main entrance to the Giardini di Castello and accommodate its attendant facilities, including ticket offices, coat checks and a police station. The design concept takes up the idea of an entrance "door" to the exhibition as a "passage", a structure connecting the different sections and places into which the exhibition areas are divided. The original design seeks to be an icon of the connection between the exhibition's various parts, rendering this connection visible. The exhibition is seen as a great container of information, which is made conveyable through the construction of a symbolic wiring system in which "The Cord" takes on proportions that can be walked through. This giant virtual network reinforces the idea of art as communication and communication as art. It is a hollow space that can convey information and visitors and their "Dreams and Conflicts" (the title of this Biennale directed by Francesco Bonami) in the various sites in which the Biennale sets the stage for its works. The piece's physical construction, in its entirety, consists of a 200-meter long steel conduit in sections 1.25 meter long, assembled in crops between 7.5 and 15 meters. In addition to the exhibition's various sites, they brought the Biennale's art works to numerous Italian art cities, including Genoa, Palermo, Turin, Lucca, Verona, Assisi, Bari and Naples. Each modular cylindrical piece has a 3-meter external radius and a 2.75 internal one. They are made by joining two 4 mm thick Corten steel sheets, which are calendered along their extrados and intrados. They are attached through two circular centers made

of rectangular sections (120x84 mm) placed at the tip of each element. The outside finish makes use of Corten steel's natural oxidization, and the internal surface is completely painted with a layer of photo-sensitive fluorescent white enamel that absorbs sunlight during the day and reflects it back with a greenish white light after sunset. On this internal surface, the titles, names and descriptions of the exhibition are impressed using dynamic lettering that seems to shift; of course, the letters are still and the movement is created by people moving through the space where they are written.

pag 524-541

**location** Giardini di Castello, Venice, Italy
**project** Entrance for the 50th Venice Biennale of Visual Arts
**client** Biennale di Venezia
**structures** Favero&Milan Ingegneria
**lighting design** Jan Van Lierde
**plan** 2003
**construction** 2003
**cost** € 600,000
**plot area** 250 sq.m
**volume** 1,200 cu.m
**contractor** Fima Cosma Silos S.r.l.

# 残障人士CDD中心

**Seregno**，意大利，2003年

本项目用地是一块位于一个托儿所后面的地块，旁边是一块将建成公园的区域，位置在一个组织混乱无序的居民区。这个功能性计划的目标是为残障人士提供服务，重点放在主要和辅助活动，教室和车间在设计构思中当作一种对残障人士处境相关的社会教育工作来进行开展。项目与周围环境构成了一种特殊的关系，坐轮椅者可以进入的公园，因为树木"种植"，"穿插"在由光滑混凝土构成的楼板上，而铺面就采用这种人造森林。这块区域因此被重新设计成为了公园的自然延续体，在其尽头处弯曲，围合建筑体量，从地面铺装到屋顶过渡自然。一个建筑正面因此呈开放性，而另一个呈封闭性，但入口处除外，这个入口可以通过一个可供车辆通行的坡道来到达。长方形的规划沿长轴分配室内各种功能空间。而这些空间通过一条有遮盖的路径与外部连接。

**页号514-523**

**地点**　Seregno，米兰，意大利
**项目**　学校
**客户**　Seregno市政府
**结构**　Matteo Fiori and Luca Varesi
**系统**　Studio Armondi - StudioTi
**规划**　2003年
**建设**　在建
**成本**　2,700,000欧元
**占地面积**　10,000平方米
**建筑面积**　1,300平方米
**体量**　4,940立方米
**承包商**　T.i.e.c.i. S.r.l.

# THE CORD

**威尼斯**，意大利，2003年

"The Cord"是第50届威尼斯双年展艺术展的装置艺术/入口。本作品定位于雕塑和建筑之间，它的正式用途是标明通向Giardini di Castello的入口并容纳其服务人员工作设施，包括售票亭、衣帽间和一间派出所。其设计概念采用了将展览入口"大门"当作一个"通道"的创意，连接展览区域内各区段和场所的构筑物。原始设计寻求成为一个展览各部分之间连接的符号，并使得这种连接为人们所看见。本次展览被视作一个极大的信息容器，信息的传达通过构建符号化的接线系统来进行，"The Cord"就是这个系统当中充当了可供人们通过的部分。这个巨大无比的虚拟网络增强了艺术作为交流手段以及交流手段作为艺术的创意。这是一个中空的空间，能够把信息，参观者及他们的"梦想与冲突"传送到双年展设置的各个展点（由Francesco Bonami指导的本届双年展的主题），也正是本届双年展为其作品设置的舞台。这件作品的实体构筑物全体由每段1.25米长的200米长的钢管构成，并装配成7.5米至15米的管束。除了通向展览的各个地点外，这件作品还将双年展的艺术作品带给了数量众多的意大利艺术城市，包括热那亚、巴勒莫、都灵、卢卡、维罗纳、阿西西、巴里和那不勒斯。每个模块化的作品圆柱体都拥有3米的外径和2.75米的内径。这些模块采用2张4毫米厚考登钢板接合制作，并沿这些钢板的拱背和内弧面进行了砑光。这些模块通过两个环形中心来固定，环形中心由长方形段（120×88 mm）制成，并放在每个元件的顶尖处。外表面层采用了考登钢的天然氧化层，内表面完全涂覆了一层光敏感型荧光白瓷漆，这种瓷漆能够在白天吸收阳光而在日落后反射出一种白绿色的光。在这个内表面上，刻印了看起来似乎会漂移的包含展览的名称和描述的动态字符；当然，这些字符本身是静止不动的，移动的效果来自于人们沿书写这些字符的空间通行的过程。

**页号524-541**

**地点**　Giardini di Castello，威尼斯，意大利
**项目**　第50届威尼斯双年展视觉艺术展入口
**客户**　Biennale di Venezia
**结构**　Favero&Milan Ingegneria
**照明设计**　Jan Van Lierde
**规划**　2003年
**建设**　2003年
**成本**　600,000欧元
**占地面积**　250平方米
**体量**　1,200立方米
**承包商**　Fima Cosma Silos S.r.l.

## EX-LAZZERI THEATRE BOOKSHOP
**Livorno, Italy, 2003-2008**

The former Lazzeri Theatre is located in downtown Livorno, in an elegant neighbourhood of the city. The renovation project aimed at creating a private multifunctional centre of culture which pivots on the main activity, namely a bookstore, also comprises restoration work and a conservative renovation. Without interfering with the identity and character of the original shell, the project exalts the distinctive traits of the composition of the interior, which is expanded through a complex work which has made it possible to obtain more space in the basement without altering the original character of the spaces. The books are treated as the leading players on an immobile stage animated by visitors and customers who are projected in a new theatrical dimension. A large ceiling lamp, made from optical fibres, descends from the ceiling covered in a reflecting steel film. The flying tower is an enormous container crossed by a sequence of galleries used for displaying books and reading; the stalls serve as a spacious room for consultation; the roof terrace is an attic-salon with a view. Exhibition and storage areas have been located in the basement while a steel and glass container on the roof houses the cafe.
**pag 542-553**

**location** Livorno, Italy
**project** Bookshop
**client** Primerose s.r.l.
**structures** Favero&Milan Ingegneria
**systems** Studio Zambonin - Studio Grava
**plan** 2003 competition, 1$^{st}$ place
**construction** 2007-2008
**cost** € 3,500,000
**built area** 1,500 sq.m
**volume** 6,932 cu.m
**contractor** Consage s.r.l.

## NEW MULTIPURPOSE CENTRE
**Trieste, Italy, 2004**

The project concerns the recovery and transformation of the 19$^{th}$ century building of the former Wine Warehouse on the waterfront of Trieste. The project does not modify the original volume, but takes at digging inside another completely independent, ethereal and translucent building, organized dimensionally from the metric of the wall piece taken from the original facade. The physical dislocation between the new "artefact" and the historic vestment allow the realization of a grand space between interior and exterior, completely covered with gilded terracotta tiles. The glass that closes the interior envelope reflects the contours of the warehouse walls and its openings allowing visibility of the activities that take place inside. The new volume is developed on four levels: the lowest level, completely underground, is used as a parking facility; the one above, at a lower level with respect to the city level, is illuminated by the space created between the original envelope and the new one; the ground floor, raised to 80cm above the street, recovers the old security level with respect to the maximum level of the tide; the new mezzanine floor, while surpassing the height of the perimeter walls, reaches the same top level of the original destroyed roof pitch.
**pag 554-559**

**location** Trieste, Italy
**project** Multipurpose centre
**client** Fondazione CRTrieste
**structures** Favero&Milan Ingegneria
**systems** StudioTi
**plan** 2004
**construction** Under construction
**cost** € 15,000.000
**built area** 3,600 sq.m
**volume** 11,000 cu.m

## VAN MELLE FACTORY RENOVATION
**Lainate, Italy, 2005-2011**

The existing industrial complex, placed in front of a large green area and a recent residential district, is part of the outskirts of Lainate. The overall reorganization of the plant comprises the construction of a new office building, warehouses and goods loading areas, new parking spaces, the redesign of the interior green areas and the spaces for handling of goods. The fragmentation of the existing volumes has been screened by an enclosure consisting of a sequence of double uprights in galvanised steel covered by perforated sheet metal onto which an "alliteration" of circular elements in glass, with variable diameter, are fixed. In addition to enclosing the "beech garden", the warehouse and the entrance building, this "wing" forms a single, "gelatinous" urban front that is more than 200 metres long. The new façade transforms the industrial complex into an ethereal urban façade which evokes both the extraordinary fountains with water jets of the Villa Litta in Lainate and the production of caramels and candies which characterizes the corporate identity. The new office building, which has not yet been built, conceived as energetically self-sufficient organism, consists of three parallel volumes connected to one another through open courts.
**pag 560-581**

**location** Lainate, Milan, Italy
**project** Offices and warehouse
**client** Perfetti Van Melle S.p.A.
**structures** Favero&Milan Ingegneria
**systems** StudioTi
**plan** 2005
**construction** 2007-2011
**cost** € 15,000,000
**built area** 12,985 sq.m
**volume** 94,571 cu.m
**contractor** Rizzani De Eccher

# EX-LAZZERI剧院书店
里窝那，意大利，2003年-2008年

这座前Lazzeri剧院位于里窝那市中心的一个优雅社区。翻修项目是一个私营多功能的文化中心，并以主要活动为支点，即其书店的实质，还包括了修复工作和保存性翻新。在不干扰原有外壳的标识和特色之下，项目发扬了内饰构造的鲜明特质，并通过一项复杂的工程对其进行了扩建，从而在地下室获得了更多的空间，而没有改变这些空间的原有特色。书籍被视作一个不活动舞台上的主演者，其动作由参观者和客户来执行，他们被纳入到新的剧院维度之下。由光学纤维制成的大型吊灯从天花板悬垂而下，上面覆盖了一层反光钢带。呈飞翔状的灯塔是一个巨大的容器，中间贯穿着一系列的走廊，走廊中陈设着书籍和读物；剧院大堂成为了一个供咨询服务使用的宽敞房间；屋顶平台成为了一个景色优美的屋顶沙龙。展示区和储藏区都位于地下室，而屋顶的咖啡屋设置在钢结构的玻璃房内。

**页号542-553**

| | |
|---|---|
| **地点** | 里窝那，意大利 |
| **项目** | 书店 |
| **客户** | Primerose s.r.l. |
| **结构** | Favero&Milan Ingegneria |
| **系统** | Studio Zambonin - Studio Grava |
| **规划** | 2003年设计竞赛，第一名 |
| **建设** | 2007年-2008年 |
| **成本** | 3,500,000欧元 |
| **建筑面积** | 1,500平方米 |
| **体量** | 6,932立方米 |
| **承包商** | Consage s.r.l. |

# 新多功能中心
的里雅斯特，意大利，2004年

项目涉及到位于的里雅斯特河岸的前葡萄酒仓库这座19世纪建筑物的修复和改造。本项目并没有改变原有的建筑体量，但通过内部深挖形成了另一个完全独立、飘逸和半透明的建筑物，其建筑形式上的节奏感来源于原有建筑正面的墙壁形式。新的"人工制品"与历史性外衣之间的实体错位让我们能够在内层与外层之间实现一个广阔的空间，并完全由涂金赤陶砖来贴面。将内部封套闭合起来的玻璃反射着仓库墙壁及其开口的轮廓线，使得我们可以看到发生在内部的各种活动。新体量的开发分成了4层：最下面的层完全位于地下，用作停车设施；上面的一层比城市标高稍低，通过原来封套与新封套之间形成的空间来照亮；地面层高于街道80cm，恢复了涨潮最高水位的旧有安全水位；新的夹层，架设在围墙高度之上，可达到与原来已损毁的屋顶斜坡相同的高度。

**页号554-559**

| | |
|---|---|
| **地点** | 的里雅斯特，意大利 |
| **项目** | 多功能中心 |
| **客户** | Fondazione CRTrieste |
| **结构** | Favero&Milan Ingegneria |
| **系统** | StudioTi |
| **规划** | 2004年 |
| **建设** | 在建 |
| **成本** | 15,000,000欧元 |
| **建筑面积** | 3,600平方米 |
| **体量** | 11,000立方米 |

# PERFETTI VAN MELLE
# 工厂翻新
Lainate，意大利，2005年-2011年

这座现有工业建筑群位于一块大面积绿地以及一个近期建成居民区的前方，属于Lainate市的郊区。这座工厂的整体改造包括了建设新办公楼、仓库和货物装载区、新停车空间、内部绿地和输送货物空间的重新设计。现有体量由一整个围墙围合，该围墙由一系列的双重装配结构的镀锌多孔板构成，多孔板上固定直径不同的玻璃圆盘。除了围起"山毛榉花园"、仓库和入口建筑物以外，这个"翼部"构成了一个长度超过200米的单一"胶状"建筑立面。这个新建筑立面将这座工业建筑群转化成了飘逸的城市立面，让人们联想起Lainate的Villa Litta所拥有的极为出色的喷射水柱喷泉，以及作为公司特色的焦糖和糖果的生产。新办公楼目前尚未建成，在设计构思上作为一种能源自给的机体，由3个平行柱体构成，柱体之间由开放庭院相连。

**页号560-581**

| | |
|---|---|
| **地点** | Lainate，米兰，意大利 |
| **项目** | 办公楼和仓库 |
| **客户** | Perfetti Van Melle S.p.A. |
| **结构** | Favero&Milan Ingegneria |
| **系统** | StudioTi |
| **规划** | 2005年 |
| **建设** | 2007年-2011年 |
| **成本** | 15,000,000欧元 |
| **建筑面积** | 12,985平方米 |
| **体量** | 94,571立方米 |
| **承包商** | Rizzani De Eccher |

# 4 EVERGREEN
**Tirana, Albania, 2005**

The city of Tirana has been the object of a general urban renewal plan for a number of years. The competition for this tower, located in the city center, came out of an earlier design by the French "Architecture Studio", winner of an international competition put on by the municipality. The proposal in the master plan involved building ten 85-meter high private towers, and ten competitions by invitation were put on to design them. The tower that Archea proposed has 6 levels of underground parking, 4 retail levels, 7 office levels, 8 residential levels and a hotel with a scenic view. The design fits into the context, taking on the lines of the surrounding city for the principle direction of its composition; the division of the project into four parts reflects the 1920s urban design by Armando Brasini based on the city's foundation on the cardus and decumanus. The tower takes a minimal amount of public space on the ground through a well-considered narrowing of the structure at the base, and symmetrically, a clear widening of the peak section, giving the total a marked sculptural effect. The movement of the tower's forms recalls that of the historic Italian landscape, merging them with deeply-rooted local traditions in the design of the façade surfaces.
**pag 582-591**

**location** Tirana, Albania
**project** Multipurpose tower
**client** Al&Gi Shipk RR - Dervish Hima
**structures** aei progetti -
Niccolò De Robertis
**systems** StudioTi
**plan** 2005 competition, 1st place
**construction** Under construction
**cost** € 25,000,000
**built area** 12,400 sq.m
**volume** 46,750 cu.m

# METROPOLITAN
**Livorno, Italy, 2005**

The set of buildings covered by the transformation project occupies a portion of the urban fabric of Livorno looking onto, through the volume of a former disused cinema, one of the busiest streets of the town center, to then develop itself toward the interior until facing, with a small residential artefact on three levels, onto a parallel road. The program envisages the construction of new commercial and residential spaces and the construction of an underground car park which develops over the entire lot. The theme is interpreted as the will to rebuild a new piece of city that can functionally link the two streets. The new urban fabric is generated through the execution of an internal open-air walkway onto which overlook the small shops and tree-lined squares connected to the parking. The crossing, the real fulcrum of the project, is designed like a notch in the compact building fabric: this aspect is expressed both spatially, through the tapering towards the top section of the path conceived as an ancient alley, and in the material sense, with Corten steel wall coverings.
The transformation of the urban sector is manifested outside the lot through new facades conceived as opaque surfaces plated in travertine blocks to construct a new-found harmony with the city.
**pag 592-599**

**with MDU architetti**
**location** Livorno, Italy
**project** Commercial and residential building
**client** Goldoncina s.r.l.
**structures** aei progetti
**systems** M&E s.r.l.
**plan** 2005
**construction** Under construction
**cost** € 7,200,000
**built area** 6,500 sq.m
**contractor** Consage s.r.l.

# NURAGIC ART MUSEUM
**Cagliari, Italy, 2006**

The project site is located on the fascinating seaside of Cagliari, in a suburban area occupied by various structures (stadium, low-income housing, fishing canal harbour). The museum described by the competition announcement has been translated in a project concept which involves architecture, town planning and landscape design. The seascape has been remodelled to give the site an identity, within which context the building acts as landmark. In fact, it consists of a monolith suspended on an artificial dune: the 7 floors of functional spaces (storage areas, laboratories, guest house, auditorium, bookshop, offices, exhibition area, panoramic roof terrace with cafe and library) pivot on a large cavea, a kind of covered square which continues below the building, accessible in every part, connected to the path along the shore. The trajectory of the latter is accentuated by routes leading to a series of collective facilities, with respect to which the museum acts as connecting void and landmark. Ascending from the sea, the dune rises to a level of +8.5 m, outlining the large square where the main entrance is located, to then descend gently towards the tree-covered parking area.
**pag 600-609**

**with Franz Prati and MDU architetti**
**location** Cagliari, Italy
**project** Museum
**client** Autonomous Region of Sardinia
**structures** Studio Chessa
**systems** Milano Progetti S.p.A.
**plan** 2006 competition, jury mention
**cost** € 36,000,000
**built area** 12,050 sq.m

# 4 EVERGREEN

地拉那，阿尔巴尼亚，2005年

地拉那市在若干年以来一直以建设更新成伟大城市计划为目标。对这座位于此城市中心的塔楼的设计竞赛方案来自于早前由法国的"Architecture Studio"事务所做出的早期设计，此设计方案正是由评论所举办的一次国际竞赛的获胜作品。总体规划当中的建议涉及建设10座85米高私营机构塔楼，并为这些塔楼的设计举行了10项设计竞赛。Archea事务所投标的塔楼拥有6层地下停车场、7层办公楼、8层住宅楼以及一个景色优美的旅馆。此设计方案相当符合本项目的背景，取围绕城市的线条作为其构成的主脉方向；将此项目区分成4个部分，反映出了由Armando Brasini根据这座城市的十字形基础而设计的1920年代的城市面貌。这座塔楼通过对地基精心地缩减，尽量少的占用了公共空间，并且对称地在楼顶部分进行了扩宽，为塔楼整体赋予一种显著的雕塑效果。塔楼形式的动感让人们回忆起意大利的历史景观，并与当地立面设计中深厚的文化传统相融合。

页号582-591

| | |
|---|---|
| **地点** | 地拉那，阿尔巴尼亚 |
| **项目** | 多功能塔楼 |
| **客户** | Al&Gi Shipk RR - Dervish Hima |
| **结构** | aei progetti - Niccolò De Robertis |
| **系统** | StudioTi |
| **规划** | 2005年设计竞赛，第一名 |
| **建设** | 在建 |
| **成本** | 25,000,000欧元 |
| **建筑面积** | 12,400平方米 |
| **体量** | 46,750立方米 |

# METROPOLITAN

Livorno，意大利，2005年

这组建筑包含在一项改造项目当中，此项目占据了里窝那城市肌理的一部分，这部分可通过废弃的前电影院的体量看到城市中心繁华的街道之一，并随后在向着内部发展自身直至通过一座3层的小住宅而面对一条并行的马路。这个计划设想了新商业和住宅空间的建设以及一个沿整个街区分布的地下停车场的建设。主题可以解读为重建一个能够从功能上将两条街道连接起来的城市新组成部分。这个新城市基本结构通过内部露天人行道的实施而形成，从这条人行道上可看到各个小商店以及连接至停车场的林荫广场。作为项目真正支点的十字路口设计就像这个紧凑建筑结构上的一个突破点：这方面在空间上通过构思一条古代小巷的道路将顶部逐渐缩小，并用考登钢墙壁贴面。这个城市区段的改造工作对地块外部做出了表现，这种表现通过镀在凝灰石上的不透明表面的新建筑立面达成，并作为一种新的建筑物与这座城市和谐统一。

页号592-599

**与MDU architetti合作**

| | |
|---|---|
| **地点** | 里窝那，意大利 |
| **项目** | 商业和住宅建筑 |
| **客户** | Goldoncina s.r.l. |
| **结构** | aei progetti |
| **系统** | M&E s.r.l. |
| **规划** | 2005年 |
| **建设** | 在建 |
| **成本** | 7,200,000欧元 |
| **建筑面积** | 6,500平方米 |
| **承包商** | Consage s.r.l. |

# 努尔吉艺术博物馆

卡利亚里，意大利，2006年

项目建址位于风光迷人的卡利亚里海滨，处于一个被各种各样的构筑物（体育馆、低收入户房屋、捕鱼航道码头）所组成的郊区。设计竞赛说明所描述的博物馆转换成一种涉及建筑设计、城镇规划和景观设计的项目概念方案。方案对海景进行了重新改造，赋予建址一种新的标识，使得这座建筑物成为周围环境当中的地标。事实上，方案由一个悬吊在人造沙丘上的巨大建筑组成：7层功能空间（仓贮区、实验室、客房、礼堂、书店、办公室、展示区、设有咖啡馆和图书馆的全景屋顶平台），一有遮盖的广场，此广场延续至建筑物之下，可从各个部分进入，并连接至海边蜿蜒的小径。这个广场被各种通向一系列公共设施的路径所强调，而博物馆对于这些路径起到了连接空地和地标的功能。地势从海岸向上提升，沙丘升高至+8.5m的标高，标明了这个大广场的轮廓，主要入口就位于这个广场上，然后略微下降通向林木掩映的停车区。

页号600-609

**与Franz Prati和MDU architetti合作**

| | |
|---|---|
| 地点 | 卡利亚里，意大利 |
| 项目 | 博物馆 |
| 客户 | Autonomous Region of Sardinia |
| 结构 | Studio Chessa |
| 系统 | Milano Progetti S.p.A. |
| 规划 | 2006年设计竞赛，评审委员会推荐 |
| 成本 | 36,000,000欧元 |
| 建筑面积 | 12,050平方米 |

# TORRE DELLE ARTI
**Milan, Italy, 2006**

The site lies in the consolidated housing row characterized by regular lots with buildings of different heights and varied facade design, albeit aligned and respecting of the consequential principle of the street fronts. The district is devoid of significant architectural and landscape elements, instead being today particularly degraded due also to a road layout now affected by a high density of vehicular traffic. The general plan concerns the definition of a new urban reference that, on the area of the pre-existing Montedison headquarters and equal in surface area with the latter, is able to accommodate residences and shops located at the foot of the building, within a context that could be an opportunity for the rehabilitation and visibility of the surrounding area. The rules on the densification of vertically developed structures and on the possibility of liberating free space on the ground come from the commission. The latter, among the proposals of the various studies consulted, has favored the solution that best characterizes the building as a dialectic expression between contemporaneity and tradition, between tradition and modernity, contrasting the desire with the need of living in Milan. The project develops infact from the image of the tower as a typology able to highlight one of the identifying characteristics of the city, developing an interpolation, or better a studied "morphing", between the varied sections of the Velasco Tower and the Pirelli skyscraper's vertical clefts. The solution proposed, therefore, is represented by the sculptural image of a vertical body divided and arranged into 4 volumetric units characterized by the presence of large intermediate grooves that accentuate the sleek silhouette beyond a lower block that

recovers the skyline of the adjacent facade. The complex, programmatically organic in architectural language, breaks the monolithic appearance of the former building wall, identifying a landmark for the entire neighborhood, a new perimeter of ground space, and a different use of the road, which enters into the lot through a small square. The project takes advantage of the collaboration with artists, architects, photographers and contemporary designers, involved in the characterization of the space. The functional allotment is structured in 4 underground parking floors (for 300 cars); ground floor with mixed use (commercial, public space and paths, access); health club and pool (4th and 5th floors); 22 floors of apartments; restaurant and roof terraces on the 23rd floor, two of which recall the Milanese tradition of hanging gardens. The interior and exterior housing spaces make up a continuous system of loggias, which provide light through large glass openings, formalizing the "double skin" technology as a criteria of energy efficiency applied to the shielding of the façades. The latter alternate with the gold-glazed ceramic stone cladding of the clefts.
**pag 610-621**

**location** Milan, Italy
**project** Residential and commercial building
**client** Babcock & Brown
**structures** Favero&Milan Ingegneria
**systems** Studio Ti
**plan** 2006-2007
**cost** € 50,000,000
**built area** 19,190 sq.m
**volume** 60,000 cu.m
**contractor** Else costruzioni S.p.A. - Pessina Costruzioni, S.p.A.

# TANGO DISCO
**Beijing, China, 2006**

The venue, one of the best known in Beijing, is located nearby the second ring-road, as the Temple of the Lamas and the Park of the Sun. The restyling of the interior (dancing floors, KTV, bars and restaurants) and the facade is aimed at creating an image reflecting the strategic, urban and social visibility of the building, at the maintenance of the structural grid supporting the old facade. The project centres on the research of a new and up-to-date urban facade capable of dialoguing both with the contemporary city and with the traditional and vernacular architecture. The new façade represents the continuity of an experimentation centring on the relationship with the surrounding tissue. This research has been conducted with a minimal expenditure of economic energies through the construction of a regular grid of aluminium panels (2x1 m) with holes of various sizes, which form a metal curtain. The surface (16x30 m) forms an interspace which serves to house the lamps. The façade system thus succeeds in interpreting the "double life" of the building: by night it is animated by a fascinating backlight; by day its appearance of compact "black box" acts as metropolitan landmark.
**pag 622-633**

**location** Beijing, China
**project** Entertainment
**client** Beijing Tango Entertainment Ltd.
**plan** 2006
**construction** 2007-2008
**cost** € 180,000
**built area** 2,000 sq.m
**façade area** 600 sq.m
**volume** 8,000 cu.m
**contractor** Beijing Hong HengJi Wall Engineering and Decoration Ltd.

# TORRE DELLE ARTI

米兰，意大利，2006 年

建址处于一排统一的房屋当中，这排房屋的特色是地块形状规则，拥有不同高度和各种立面设计的建筑物，整齐划一且遵循街道立面的规律。但是这个街区缺乏有重要意义的建筑和景观元素，而且如今被一条高密度车流所影响的道路布局而显得特别退化。总体规划考虑到在以前存在的Montedison总部想在另一块地块上建造一个相同面积新的城市参照物，底层容纳住宅与商店，并且有机会提高周边区域的受关注程度。有关增加垂直发展构筑物密度以及解放地面自由空间可能性的规则都由委员会来制订。在咨询的各个建筑事务所的建议书中，委员会偏向于采用能够最好地体现出这座建筑物作为当代化与传统、传统与现代性之间辩证特征的解决方案，这点与想要满足在米兰生活需要的欲望相抵触。这个项目方案实际上由塔楼作为一种能够突出城市标识特点的象征性形象发展而来，在Velasco塔楼的变化部分与Pirelli摩天大楼垂直裂口之间形成一种插入形式或更好的一种深思熟虑的"变形"。因此，所建议的解决方案是由分割并布置成4个体量单元的雕刻式形象来代表，这些体量单元的特色是大型中间沟槽的存在，强调了超出较低矮街区的流畅轮廓，并恢复了相邻建筑正面的空中轮廓线。这座建筑群在建筑语言上具有纲领性的有机特色，打破了原有建筑物墙壁的单调外观，为整个邻近地区树立了一座地标，为地面空间提供了新的周边，而且对道路进行了不同的使用，让道路经由一个小广场后再进入这个地块。这个项目利用了与艺术家、建筑设计师、摄影师和现代派设计师合作的优势，参与到了空间的特性表达当中。功能分配方式在结构上分为4层地下停车场（300辆汽车）；混合用途底层（商业、公共空间和道路、入口）；健身俱乐部和游泳池（第4层和第5层）；22层的公寓；第23层的餐馆和屋顶平台；其中两处会让人回想起米兰的悬园传统。内部和外部房屋空间组成了一套连续的门廊系统，提供了照明光线的大型玻璃开口，形成了"双层蒙皮"技术作为应用于建筑正面屏蔽层的能源效率标准。后一项技术与裂口的镀金瓷砖砖覆层交替使用。

**页号610-621**

| | |
|---|---|
| **地点** | 米兰，意大利 |
| **项目** | 住宅和商业楼 |
| **客户** | Babcock & Brown |
| **结构** | Favero&Milan Ingegneria |
| **系统** | Studio Ti |
| **规划** | 2006年-2007年 |
| **成本** | 50,000,000欧元 |
| **建筑面积** | 19,190平方米 |
| **体量** | 60,000立方米 |
| **承包商** | Else costruzioni S.p.A. - Pessina Costruzioni, S.p.A. |

# 糖果俱乐部

北京，中国，2006 年

作为北京最著名的娱乐场所之一，糖果俱乐部的位置靠近二环路，位于雍和宫与地坛公园之间。内部装饰（舞池、KTV、酒吧和餐馆）和建筑立面的重新设计旨在创造一种能够反映这幢建筑物的策略、城市和社会显著性的形象，并对支撑旧建筑正面的框架进行维护。项目重点放在研究一种能够同时与现代都市和传统本地建筑对话的新型且最时尚的城市建筑立面。新的建筑立面代表了以与周边城市肌理关系为中心的试验探索的继续。这项研究采用了经济能源方面支出最少的方式展开，通过建造形状规则且上布各孔径孔洞的铝板（2×1米）框架来构成一个金属帷幕。这个表面（16×30米）构成了一个功能用于容纳照明灯的空间。建筑立面系统因此成功地解释了这座建筑物的"双重生活"：在夜晚由绚烂迷人的背景灯光赋予其十足动感；在白天呈现作为大都市地标的结构紧凑的"黑盒子"外观。

**页号622-633**

| | |
|---|---|
| **地点** | 北京，中国 |
| **项目** | 娱乐场所 |
| **客户** | 北京糖果娱乐有限公司 |
| **规划** | 2006年 |
| **建设** | 2007年-2008年 |
| **成本** | 180,000欧元 |
| **建筑面积** | 2,000平方米 |
| **建筑正面面积** | 600平方米 |
| **体量** | 8,000立方米 |
| **承包商** | Beijing Hong HengJi Wall Engineering and Decoration Ltd. |

## MUNICIPALITY OFFICES AND CULTURAL CENTER
### Figline Valdarno, Italy, 2006

The project, located in the southern part of the walled city, consists of the transformation of an old school building from the early years of the Twentieth century, which is involved in a general urban upgrading. The plan centres on the creation of a multifunctional centre: library, archive, museum of the ancient Spezieria Serristori, the municipal offices and thus the Town Hall. The new building body retains the old C-shaped plan, accentuated by the extension of the two existing wings; but the architectural aggregate now plays a different urban role, thanks to the addition of a large projecting roof – which covers the entrances from the street and from the park behind the building – and a tower which, recomposing the alignment of the façade, becomes a new urban landmark. The functional distribution clearly distinguishes the two parts, the areas housing administrative offices – located on the first floor and in the tower – and the public ones, which are conceived as an extension of the ground floor, to the squares sheltered by the new projecting roof, which may house temporary events or shows. The entire aggregate is rendered homogeneous by the facing in natural stone which echoes, in a contemporary key, the constructive and material characters of the surroundings.
**pag 634-645**

**location** Figline Valdarno, Florence, Italy
**project** Municipality offices and cultural center
**client** Comune di Figline Valdarno
**structures** Studio G.T.A.
**systems** M&E S.r.l.
**plan** 2006
**construction** Under construction
**cost** € 4,000,000
**built area** 3,010 sq.m
**contractor** CFC costruzioni S.r.l.

## FIRENZE NOVA RESIDENTIAL COMPLEX
### Florence, Italy, 2006-2010

The former touristic structure located in Firenze Nova, is configured like a huge seven-storey artefact extended along a wide street in the northwest of the city. The property was acquired from a real estate society with the intention of transforming it into a residential building with prestigious but small homes: 145 including studios and one-bedroom with the exception of the corner area with apartments equipped with two bedrooms. The new subdivision was obtained maintaining the structure unaltered, but remodeling the dividing elements and the facade tops to transform the preceding residence, developed on two levels, into single-level homes. The building has been covered with a steel carrying structure to realize, along the perimeter, a system of loggias shielded by a series of perforated panels constituted of white glazed ceramic elements and mounted on steel frames. The design of the ceramic elements has been studied with the artist Franco Ionda: an irregular figure that reproduces an asymmetric texture. At the ground floor the new commercial spaces are characterized by a completely glazed facading that makes the complex's base transparent.
**pag 646-653**

**location** Florence, Italy
**project** Residential complex
**client** S.I.IMM. Firenze S.r.l.
**structures** Vega Ingegneria - Lorenzo Checcucci
**systems** Studio Carbone
**plan** 2006-2007
**construction** 2007-2010
**cost** € 13,234,000
**plot area** 3,300 sq.m
**built area** 14,340 sq.m
**parking area** 1,805 sq.m
**volume** 40,460 cu.m
**contractor** Q5 S.r.l.

## UBPA B3-2 PAVILION WORLD EXPO 2010
### Shanghai, China, 2007-2010

To meet the express wishes of the client, the building is a simple rectangular box, 78x28 meters, completely free inside to create a neutral space that can accommodate the exhibits of the cities participating in the event. As the project is part of a cooperation program between the Expo and the Italian Ministry of the Environment, the design also turned the concept of an industrial-inspired envelope into a mechanism for spreading natural light so that the space could be fully lit without using any energy. To these ends, the roof was designed as a shed structure along the building's long side, crossed by steel beams clad so as to form reflecting surfaces that spread the light from above. The design was also conceived to allow for the building's complete conversion and reuse. It was designed and built to be able to be disassembled and reassembled in another place. For this reason, the entire building was made with mortarless technologies that let over 90% of the parts used in the construction be recovered. The outside walls are a metal structure clad with silicon textile panels to turn the building container into a soft, vibrating surface.
**pag 654-673**

**location** Shanghai, China
**project** Exhibition pavilion
**client** World Expo Shanghai 2010 Holding Company
**structures** Favero&Milan Ingegneria
**systems** Favero&Milan Ingegneria
**plan** 2007
**construction** 2007-2010
**cost** € 2,000,000
**plot area** 3,000 sq.m
**built area** 2,000 sq.m
**contractor** Shanghai Construction Company

## 市政办公楼和文化中心
**Figline Valdarno，意大利，2006年**

本项目位于一座城墙环绕的城市的南部，由20世纪早期的旧学校建筑的改造工程，并涉及到总体城市升级工作。规划方案以创建多功能中心为重点：图书馆、档案馆、古代Spezieria Serristori博物馆、市政办公楼和市政厅。新建筑主体保留了旧有的C形平面布置，并对两座现有翼部的延伸进行了强调；但如今这个建筑集合体执行着不同的城市职能，而由于添建了大型外伸屋顶，这个屋顶覆盖了从街道方向以及从建筑物后面公园方向的入口，以及加建一座与建筑正面对齐的塔楼，使之成为了新的城市地标。功能性分布清晰地分出了这两个部分，位于第一层以及塔楼之内用于容纳行政管理办公室的区域和公共区域，从底层的延伸直至由新建外延屋顶所遮蔽的广场，用于举行临时性的活动或表演。整个建筑集合体通过采用天然石材的表面形成了均质化的效果，而这样的表面以流行的基调反映周围建筑的构造和材料特色。
**页号634-645**

**地点** Figline Valdarno，佛罗伦萨，意大利
**项目** 市政府办公楼和文化中心
**客户** Comune di Figline Valdarno
**结构** Studio G.T.A.
**系统** M&E S.r.l.
**规划** 2006年
**建设** 在建
**成本** 4,000,000欧元
**建筑面积** 3,010平方米
**承包商** CFC costruzioni S.r.l.

## FIRENZE NOVA
## 居住建筑群

**佛罗伦萨，意大利，2006年-2010年**

这座位于Firenze Nova的前旅游性构筑物，在布局上是一个沿城市西北方向的宽阔街道延伸的7层建筑。此物业从一家房地产社团购得，其目的是将其改造成一栋住宅楼，设置优雅紧凑的小型公寓：145户，包括单室公寓和双室公寓，除局部区域设置的两卧室公寓例外。这种新的住宅分隔是在保持结构不改变的情况下做到的，但对分隔元素以及建筑正面顶部进行了结构改变，以便将以前开发成2层的住宅改造成单层公寓。沿着建筑物的边缘设置钢结构框架的回廊，安置在框架上的百叶为一系列多孔面板，这种面板由白色的上釉瓷制元素构成并安排在钢框架之上。这种瓷制元素的设计通过联合艺术家Franco Ionda进行了研究：一种体现了不对称纹理的不规则图形。在底层，新商业空间的特色体现在完全玻璃化的建筑正面，让这座建筑的底层呈现透明化。
**页号646-653**

**地点** 佛罗伦萨，意大利
**项目** 住宅建筑群
**客户** S.I.IMM. Firenze S.r.l.
**结构** Vega Ingegneria - Lorenzo Checcucci
**系统** Studio Carbone
**规划** 2006年-2007年
**建设** 2007年-2010年
**成本** 13,234,000欧元
**占地面积** 3,300平方米
**建筑面积** 14,340平方米
**停车面积** 1,805平方米
**体量** 40,460立方米
**承包商** Q5 S.r.l.

## 2010世博会UBPA
## B3-2展厅

**上海，中国，2007年-2010年**

为了满足客户快速建造的希望，方案上这座建筑物由一座简单的长方形盒体构成，78×28米，内部完全空置，可以构建一个用于容纳参与这次展会的各个城市展品的空间。由于本项目是世博会与意大利环境部之间的合作计划的组成部分，其设计方案也将灵感来源于工业的封套概念转化成一个用于散布自然光线的机械装置，让这个空间可以在不使用任何能源的情况下完全明亮。最后，屋顶设计为一个沿这座建筑物长轴延伸的棚式构造，由钢梁承重，吊顶构成纵向向下漫射光照的反射表面。此项设计还在构思上允许这座建筑物进行完全的转换和重复利用。这座建筑物设计和建造成了一种可以拆卸后在另一个地方重组的形式。因此，整个建筑物采用不用灰浆的建筑技术进行建造，建造过程中用到的部件有90%以上可以回收。外墙采用了一种金属框架的，镶硅胶纺织材料的面板，用于将建筑物的容器转化成一种柔软、有振动感的表面。
**页号654-673**

**地点** 上海，中国
**项目** 展馆
**客户** World Expo Shanghai 2010 Holding Company
**结构** Favero&Milan Ingegneria
**系统** Favero&Milan Ingegneria
**规划** 2007年
**建设** 2007年-2010年
**成本** 2,000,000欧元
**占地面积** 3,000平方米
**建筑面积** 2,000平方米
**承包商** Shanghai Construction Company

## EX MANIFATTURA TABACCHI
**Cava dei Tirreni, Italy, 2007**

The old headquarters of the tobacco factory of Cava dei Tirreni vaunt a singular location, both because it is found in the centre of town and due to the role the building plays in the historical memory of the city: originally a stately mansion, it has been converted into a convent-orphanage, to then become the symbol of the local economy, before being abandoned. The project aimed to make the ground floor of the entire complex available to the public through a program of entertainment, shops and leisure activities, while the residential areas are located on the upper floors, conceived so as to reconnect numerous urban spaces. The original volume, distributed around a court, has been turned into a hotel structure, while the demolition of the industrial sheds behind provides space for underground parking areas and the building above, with dwellings and shops. The complex plan and sections are studied so as to offer a number of different glimpses and views which carry the eye beyond the limits of the project as such, thus recreating a new vision of the city. The recovery of every free area has been ideated so as to form a sequence of footpaths, widening, courts and gardens which expands the borders of the accessible parts of the old town. The materials chosen for the facades reflect the old manufacturing tradition linked to artistic ceramics and glazed earthenware.
**pag 674-687**

**location** Cava dei Tirreni, Salerno, Italy
**project** Residential and hospitality complex
**client** Manifatture Sigaro Toscano S.r.l.
**structures** Favero&Milan Ingegneria
**systems** StudioTi
**plan** 2007
**construction** In progress
**plot area** 13,482 sq.m
**built area** 13,310 sq.m

## PARK OF MUSIC AND CULTURE
**Florence, Italy, 2007**

The railway area of Firenze Porta al Prato is delimited on one side by the former Leopolda Station (location of cultural events) and by the underground station – currently under way – and on the other, by the Parco delle Cascine. The vicinity of the current Municipal Theatre (500 m), the hinge between the historical centre and territorial road network, and the Park's leisure activities, determine the both cultural and playful vocation which the competition proposes to improve with the construction of the new head offices of the Florentine Maggio Musicale. The project is to include: a lyrical theatre to seat 2,000 people, an auditorium with 1,000 seats, including an open-air cavea, technical stage-set workshops, commercial activities with relative services and public spaces. The project tackles the significance of the architectural object not just in terms of structure and function, but also in terms of art and city life: the works of the master Giuliano Vangi show how architecture and sculpture are perceived as elements of a single space, as in an ancient, Renaissance town; and the architectural symbols of perception and qualification derive from the study of how the area is visually approached. The required functions are articulated in a fluid, sculptural form, which identifies each urban rapport in a plastic continuum of projections, indentations and variations in level: it is not structured through hierarchical views, but through focal points and cross axes, presenting itself with conformations and perspective links, and a variety of functions.
The line of approach, along which the access to the underground car park and pedestrian underpass are located, coincides with the access way to the historical centre and with the road network, including that

of public transport. It becomes a both functional and formal axis: shops, eateries and leisure centres lure the visitor along the architectural promenade towards the foyer and music spaces. The latter consist of two cantilevered foreparts, which cover respectively the terraces of the open-air cavea – projection of the concert hall – and part of the square beneath the auditorium. The elements develop in a crescendo, which culminates in the 40 m-high scene tower. The façade of building, articulated in such a way, overlooking 300 m of public urban space, contains a much sought-after perceptive value within. The stone blocks, enhanced by the whitish-green enamelled terracotta cladding, a historic memory of the Florentine Romanesque period and of the Tuscan ceramics tradition, are opened up with glazed surfaces, through which the building gradually reveals its internal articulations.
**pag 688-709**

**location** Florence, Italy
**project** Theatre and auditorium
**client** Presidenza del Consiglio dei Ministri
**structures** Favero&Milan Ingegneria
**systems** Studio TI
**plan** 2007 competition, 2nd place
**cost** € 80,000,000
**built area** 60,000 sq.m
**contractor** Gia.Fi. Costruzioni S.p.A.

# 前制烟厂

**Cava dei Tirreni，意大利，2007年**

# 音乐文化公园

**佛罗伦萨，意大利，2007年**

这座Cava dei Tirreni卷烟厂的旧总部可称得上处于一个独一无二的地点，这既是因为它处于城市的中心，也是因为这座建筑物在这座城市历史记忆中所发挥的作用：最初是一幢宏伟的豪宅，后来被改造成一个修道院和孤儿院，之后又变成了当地的经济符号，最终被废弃。项目的目标是将整个建筑群的底层通过一项娱乐、商店和休闲活动开发计划而提供给公众使用，同时在上层部分设立住宅区，从构思上重新连通众多的城市空间。原来的体量环绕一个庭院分布，如今改造成了一个酒店，同时通过拆除后部的工业棚式建筑提供了地下停车场的使用空间以及用于地上包含住宅和商店的建筑物的空间。这座建筑群的规划和各个部分均经过了研究学习，提供了一系列不同的观察和观赏机会，将人们的注意力引至项目界限以外，从而重新创造了这座城市的一个新风景。每一个闲置区域的重塑都经过了构想，以形成一系列的人行道、广场、庭院和花园，将边界拓展到旧城镇的可进入部分。为建筑正面所选择的材质反映了与艺术瓷器和上釉陶器相关联的古老制造业传统。

**页号674-687**

| | |
|---|---|
| **地点** | Cava dei Tirreni，萨勒莫，意大利 |
| **项目** | 住宅和酒店建筑群 |
| **客户** | Manifatture Sigaro Toscano S.r.l. |
| **结构** | Favero&Milan Ingegneria |
| **系统** | StudioTi |
| **规划** | 2007年 |
| **建设** | 在建 |
| **占地面积** | 13,482平方米 |
| **建筑面积** | 13,310平方米 |

佛罗伦萨Porta al Prato铁路区，其一侧挨着以前的Leopolda Station（文化活动地点）和正在兴建中地铁站，另一侧靠着Parco delle Cascine。此区域毗邻当前的市剧院（500米），并作为历史性中心和地方路网之前的连接点，而且此公园的休闲活动决定了本次设计竞赛所提出的文化和游玩度假的概念将随着Florentine Maggio Musicale新部的建设而得到提升。本项目包括：一座能容纳2000人的歌剧院、一座有1000个座位的礼堂、一个露天阶梯、技术舞台布景工作间、拥有相关服务和公共空间的商业活动场所。项目不仅从结构的方面，还从艺术和城市生活的方面，满足了此建筑设计目标的重要性：Giuliano Vangi大师的作品反映了建筑设计和雕塑如何作为一个单独空间的元素被人感受的，就象文艺复兴时期的城镇一样；以及来自于此区域如何从视觉上被看待的研究工作的感知和资格的建筑设计符号。所要求的功能以一种流畅、雕塑般的形式连接起来，通过凸凹起伏以及标高的变化所形成的塑性连续统一体对每一个城市融洽关系做出了标识：它并没有通过层次化视图进行构造，而是通过焦点和横轴，用形态和透视链接以及各种各样的功能来表达自身。进出线路，以及随之布置的地下停车场和行人地下通道的入口，在位置上与通向历史性中心以及与路网之间包括公共交通设施的进入通道相重合。这就演变成了一条功能上和形式上的中轴线：商店、小餐馆和休闲中心激发了参观者沿着这个建筑设计的散步道路前往剧场门厅和音乐空间。后者由两个悬臂式的前段构成，这两个前段又分别覆盖着露天梯阶的平台 – 音乐厅的凸出部分 – 并且是礼堂下方广场的组成部分。这个元素以一种渐强的方式发展，并在40米高的景观塔处达到最高点。建筑正面相互连贯的方式让人们可以俯瞰300米的公共城市空间，包含这人们所探求的审美价值。而石建街区，

在白绿相间的赤陶贴面的增强之下，作为佛罗伦萨罗马式建筑时期以及托斯卡纳瓷器传统的历史记忆，建筑内部格局通过玻璃幕墙部分展现。

**页号688-709**

| | |
|---|---|
| **地点** | 佛罗伦萨，意大利 |
| **项目** | 剧院和礼堂 |
| **客户** | Presidenza del Consiglio dei Ministri |
| **结构** | Favero&Milan Ingegneria |
| **系统** | Studio TI |
| **规划** | 2007年设计竞赛，第2名 |
| **成本** | 800,000,000欧元 |
| **建筑面积** | 60,000平方米 |
| **承包商** | Gia.Fi. Costruzioni S.p.a. |

## MILANOFIORI 2000
**Assago, Italy, 2007**

The design of a building in the centre of the Milanofiori 2000 area is based on the indications of the master plan prepared by EEA Architects, in which a selection of Italian firms have been asked to participate. The complexity of the project lies in the particular conditions of the plot which is "embedded" between the high level of the artificial ground – which covers the parking areas below, thus creating the new central urban square – and the lower level of the park. The building also acts as backdrop, barrier between the residential area and the shopping area and takes the form of a tall urban front capable of collecting and containing the visual sphere of the public space and the collective life. The density and overall volume of the project is therefore "necessary" even if it proves to be decidedly problematic due to the proportions and overall dimensions. This difficulty has been interpreted as a key of interpretation of the project, which has been re-modulated and "scaled" through a subdivision into two lower units, one with three floors and another with six, superimposed but somewhat shifted. The building is consequently coherent with the tall office blocks with ten or eleven floors designed by Van Egeraat, while the fragmentation make the building reacquire, architecturally speaking, a dimension that is more proportionate to the park and the buildings in front of it.
**pag 710-719**

**location** Assago, Milan, Italy
**project** Residential and commercial
**client** Milanofiori 2000 S.r.l.
**structures** Intertecno
**systems** Intertecno
**plan** 2007
**construction** In progress
**cost** € 29,000,000
**buit area** 24,848 sq.m

## GEL - GREEN ENERGY LABORATORY
**Shanghai, China, 2008**

The project, located inside the Minhang Campus of Jiao Tong University, arose from a collaboration between the university and the Italian Ministry of Environment and Protection of Land and Sea, for the realization of a center for the research and dissemination of low environmental impact building technologies. The building, named GEL (Green Energy Laboratory) is designed as a simple and compact body endowed with a central courtyard covered by an ample skylight. This space, surrounded by walkway balconies, creates a gap capable of energy consumption optimization since it functions, during sunny winter days, as a heat collector, in the summer period like a ventilation duct for the warm air that it produces internally. The first two levels of the building house laboratories, conference and control rooms, classrooms and exhibition spaces; the third and top floor includes two representative apartments enclosed in the shape of a "house" covered by a pitched roof made from photovoltaic panels. Every single environment enjoys the benefits derived from the maximization of ventilation and natural lighting through a dual-facing configuration: towards the court and towards the exterior where a double skin guarantees the control and the screening of sun rays on the glazed surfaces.
**pag 720-725**

**location** Minhang Campus of Jiao Tong University, Shanghai, China
**project** Center for research
**client** Jiao Tong University, Shanghai
**structures** Favero&Milan Ingegneria
**systems** TIFS Ingegneria
**plan** 2008
**construction** In progress
**plot area** 1,500 sq.m
**built area** 4,850 sq.m
**volume** 27,000 cu.m

## ARCHITECTURE DEPARTMENT BUILDING
**Tripoli, Libya, 2008**

In the area, south-east of Tripoli, an orthogonal grid inspired by the model of American campuses has been developed on the typical plains covered by olive trees that border on the desert. The plot reserved for the architecture faculty lies alongside that of the new library, central element in a project aimed at renovation and expansion. The concept rejects a hierarchical arrangement of the façades, while the references to local architectures with their closed façades which protect from the sun and the winds, look to the interior courts; this is a pretext to play with the relationship between interior and exterior, public and private. Four volumes are arranged to form a four-leaf clover around a covered central square, crossed by the two orthogonal axes which identify the public paths. The entrances are hardly visible in the fluid continuum of the facades without sharp corners, as if modelled by the wind, and of the surfaces that are directly exposed to the sun. The external façade features a continuous system of structural steelwork onto which gold-glazed earthenware tiles are fixed: a union between the passive technology of the sunscreen and the characteristic patterns of traditional wooden grating screens and mosques. The protective exterior is contrasted by the interior elevations, which are completely glazed, showing the students arranged in rings on the three floors.
**pag 726-735**

**location** Tripoli, Libya
**project** Public building
**client** Odac – Meftah Waggah
**local consultants** N.C.B., Mustafà Mezughi, Mohamed Gheblawi
**plan** 2008
**cost** € 15,000,000
**built area** 8,000 sq.m
**volume** 30,000 cu.m

# MILANOFIORI 2000
**Assago，意大利，2007年**

在Milanofiori 2000区域中心设计的这座建筑物以EEA Architects所编制的总体规划的指示为基础展开，这个总体规划还邀请了一些有实力的意大利建筑事务所参与其中。这个项目的复杂性在于这一地块的特别条件，它"嵌入"在标高较高人造地面，此地面覆盖着下面的停车场并形成了一个新的中央城市广场，与标高较低的公园之间。这座建筑物还充当着背景以及住宅区与商店区之间的屏障，并采取了高于城市正面的形式，能够汇集和包容公共空间以及集体生活的视野。因此，即使由于比例和整体尺度的原因而证明其存在问题，但是这个项目的密度和总体量仍然是"必要的"。这种困难一直被作为此项目处理的关键点，从通过划分成两个较低单元来进行了重新调整和"缩放"，一个为3层，另一个为6层，成阶层状但有一些偏移。这座建筑物也随后与高耸的办公楼部分保持了一致，这座10层或11层的办公楼由Van Egeraat设计，从建筑设计上说，一种与公园以及公园前建筑比例更加适当的尺度。

**页号710-719**

**地点**　Assago，米兰，意大利
**项目**　住宅和商业建筑
**客户**　Milanofiori 2000 S.r.l.
**结构**　Intertecno
**系统**　Intertecno
**规划**　2007年
**建设**　在建
**成本**　29,000,000欧元
**建筑面积**　24,848平方米

# GEL – 绿色能源实验室
**上海，中国，2008年**

这个项目的位置在上海交通大学阂行校区，其起源是这所大学和意大利环境与国土海洋保护部的合作而来，用于研究环境保护和建筑协调技术的中心。这座建筑物的名称为GEL（绿色能源实验室）设计作为一个简单紧凑的主体，并包含一个被天窗所覆盖的中央庭院。这个空间被走道回廊所环绕，形成了能够优化能源消耗量的间隙，它在阳光良好的冬日可作为一个热量采集器来发挥作用，在夏天时段好比内部热空气的排风管道。这座建筑物的头两层容纳着实验室、会议室和控制室、教室以及展示空间；第3层和顶层包括了两个代表公寓，它们被包围在由光伏电池板构成的斜屋顶所覆盖的"房屋"形状之内。每一个单独的环境都享受着通过双面向配置方式获得的通风和自然照明的好处：一面朝向庭院以及一面朝向外部，而且双层幕墙也保证了对阳光直射的控制效果。

**页号720-725**

**地点**　上海交通大学阂行校区，上海，中国
**项目**　研究中心
**客户**　上海交通大学
**结构**　Favero&Milan Ingegneria
**系统**　TIFS Ingegneria
**规划**　2008年
**建设**　在建
**占地面积**　1,500平方米
**建筑面积**　4,850平方米
**体量**　27,000立方米

# 建筑系楼
**的黎波里，利比亚，2008年**

在这块位于黎波里东南部，并被油橄榄树所覆盖的处于沙漠边界的典型平原上，受美国校园模式的启发，已经开发了一种正交网格结构的校园。这个保留供建筑系使用的地块位于新图书馆旁边，这个项目的重点是翻新和扩建。概念设计舍弃了建筑正面层次化排布的方式，同时参考当地建筑采用封闭建筑正面而面朝内部庭院的设计方式以防止阳光曝晒和风沙侵袭困扰的特点；而且还可以次处理好内部与外部、公共和私密之间关系。4个体量经安排构成了一个围绕着有顶盖中央广场的4叶首蓿的外形，广场当中布置有两个正交中轴线，用于标识公共道路。在没有尖锐角部的建筑正面就像随风成形的流畅连续体，其当中几乎看不到入口，那些直接暴露在阳光照射下的表面也是如此。外部建筑正面的特色是一套结构钢工程的连续系统，这套系统上固定着金色釉面的陶砖：一种被动遮阳技术与传统木制栅栏和清真寺特色图案的联合体。保护性外饰层与内部立面形成鲜明对比，内部立面完全玻璃化，显示出在这3层楼内成通透的环形排列的学生空间。

**页号726-735**

**地点**　的黎波里，利比亚
**项目**　公共建筑
**客户**　Odac – Meftah Waggah
**当地咨询公司**　N.C.B., Mustafà Mezughi, Mohamed Gheblawi
**规划**　2008年
**成本**　15,000,000欧元
**建筑面积**　8,000平方米
**体量**　30,000立方米

# KPM TOWER
**Dubai, United Arab Emirates, 2009**

The project is located in the scenario of the Dubai Marina, between the sea and the artificial mirrors of Jumira Lake Tower. It centres on the architectural definition of the tower signed by Katherine Price Mondadori, which has already been the subject of considerable media attention.

The purpose has not been to build a skyscraper among skyscrapers, but an urban landmark, an assertion of the sense of belonging to the Arab world and at the same time to a cosmopolitan, innovative and unique city.

The building stands out from the consolidated building tissue, featuring a hypermodern typology which is nevertheless permeated by references to the local culture: a sense of protection from the sun, inspired by the traditional wooden grating screens, is evoked by the pattern of lights and shades on the elevations, which are massive yet transparent.

Dwellings, shops, services (gym, swimming pool, cafe, Turkish bath, doorkeeper, heliport) are attuned to a register of exclusivity and studied elegance which goes beyond a foregone opulence, featuring a new formula of participative sharing of the environment as a whole by inhabitants and managers.

In a metropolis where the car dominates, the new tower faces The Walk, one of the few extraordinary promenades of the city; in a place where the facades mirror one another, it stands out by virtue of its ability to represent itself, with an own sophisticated image, in a country where one pays little attention to the energy issue, as it is economically irrelevant, it features a shell characterized by a low environmental impact and limited management and maintenance.
**pag 736-749**

**location** Dubai, United Arab Emirates
**project** Residential and commercial complex
**client** Marina Exclusive L.t.d.
**structures** Sinergo
**structures consultant** aei progetti - Niccolò De Robertis
**systems** Sinergo
**plan** 2009-2010
**cost** € 60,000,000
**plot area** 3,382 sq.m
**built area** 29,800 sq.m
**volume** 110,000 cu.m
**contractor** GTCC

# ART CUBE
**Casalbeltrame, Italy, 2009**

In the farm landscape of Casalbeltrame, an old farmhouse, tucked into the vast stretches of rice fields typical of the Piedmont plains, was turned into an art center. The place is a workshop where artists share both their creative work and their daily lives. After the building was remodeled and adapted to its new function, a request was made for a volume in the center of a large square courtyard. It was meant to mark the space for working on marble outdoors and be a lightweight cover for working in the summer while conveying a symbolic value in defining a personality for the large empty space. The space has square plan on which a frame of corten steel columns stands, defining a cubic space 9 meters of side. From the square form of the roof, a series of circular bars, also Corten steel, converges towards the tip of a strut that projects diagonally up to 16 meters. On the cables that support the large diagonal piece against the cube's top peak, a microperforated PVC canvas can be stretched to "construct" a large shaded area underneath. In this way, the installation forms a large sundial in the square, measuring time and the passing of the hours by throwing its shadow on the ground.
**pag 750-755**

**location** Casalbeltrame, Novara, Italy
**project** Sculpture pavilion
**client** Materima S.r.l.
**structures** Map Carpenteria Metallica S.r.l.
**plan** 2009
**construction** 2009
**cost** € 157,000
**surface** 81 sq.m

# KPM 摩天楼

杜拜，阿拉伯联合酋长国，2009年

本项目位于迪拜海岸线，座位在海洋与Jumira Lake Tower高层区之间。由Katherine Price Mondadori所设计的建筑形式，已吸引相当的媒体关注。项目目标不是建设一座摩天楼之中的摩天楼，而是一个城市地标，一项属于阿拉伯世界的归属感的宣言，并且同时是一座世界性、充满创新和独一无二城市的象征。这座建筑物从统一的建筑基本结构当中脱颖而出，体现出一种超现代象征主义的特色，但也渗透了对当地文化的参照：一种防护阳光曝晒的感觉，灵感来源于传统的木制格栅遮阳蓬，并在立面上形成明暗实虚的光影效果，使建筑体量既厚重又透明。 住所、商店、服务设施（健身房、游泳池、咖啡馆、土耳其浴室、门房、直升机坪）都注入了一系列独特性和精心安排的优雅风格，并超越了过去的富贵辉煌，特色体现在由居住者和管理者作为整体一部分参与和共享环境的新公式。在汽车占主导地位的大都市，这座新建塔楼面对着"The Walk"，这座城市少数几项极为非凡的散步场所之一；在这样一个建筑玻璃幕墙相互映照的场所，该大楼以其与众不同的沙漠金色的v型预制水泥板的百叶系统来表现自身的与众不同，并且在这样一个在经济上没有能源匮乏问题的国家，其特色在于其建筑外壳对环境的低影响及低管理维护成本。

**页号736-749**

| | |
|---|---|
| **地点** | 杜拜，阿拉伯联合酋长国 |
| **项目** | 住宅和商业建筑群 |
| **客户** | Marina Exclusive L.t.d. |
| **结构** | Sinergo |
| **结构咨询机构** | aei progetti - Niccolò De Robertis |
| **系统** | Sinergo |
| **规划** | 2009年-2010年 |
| **成本** | 60,000,000欧元 |
| **占地面积** | 3,382平方米 |
| **建筑面积** | 29,800平方米 |
| **体量** | 110,000立方米 |
| **承包商** | GTCC |

# 艺术方块

Casalbeltrame，意大利，2009年

在Casalbeltrame的农田景观当中，一座藏在广阔的Piedmont平原典型的稻田之内的旧农舍经改造变成了一个艺术中心。这个场所是艺术家们分享自己创作作品以及日常生活的工作室。在这座建筑物重新改造并适应其新功能之后，对位于大型庭院中心的一部分体量提出了一项要求。这项要求的目的是标出用于从事室外大理石创作的空间，并设置轻重量的顶盖以便能在夏天工作，同时传达一种用于定义这个大型空闲空间个性的符号价值。这个空间正方形布局，并构建了一个考登钢柱基座框架，界定了一个各边为9米的立方体空间。从框架顶部支出一系列圆弧型的考登钢条，沿对角线伸展，最长达16米。在支撑着大型对角线部件的钢缆上，拉伸一个微孔型PVC帆布来"构建"一个下方的大型荫▌区域。通过这种方式，此装置在广场内构成了一个大型日规，能够通过对地面的投影来表示钟点和计时。

**页号750-755**

| | |
|---|---|
| **地点** | Casalbeltrame，诺瓦腊，意大利 |
| **项目** | 雕塑展厅 |
| **客户** | Materima S.r.l. |
| **结构** | Map Carpenteria Metallica S.r.l. |
| **规划** | 2009年 |
| **建设** | 2009年 |
| **成本** | 157,000欧元 |
| **表面积** | 81平方米 |

## MADAME DAI CULTURE AND ART CENTER
Changsha, China 2011

The project is located in the north-eastern area of the master plan by the KPF firm of New York for the new urbanization of the territory by the Meixi Lake, a more than 40 hectares large artificial lake located in the Daheexi district. Conceived as a pilot project for the sustainable development of the Changsha area, the new Culture Centre fits harmoniously into the organic design of the town plan, in accordance with the concept of the project, which relates the aspects of sustainability, integration between the human intervention and the natural landscape, and references to the cultural identity of the place. The complex contains many functions: a theatre, an art gallery and an auditorium - united in a single volume - and a shopping centre, a hotel, offices and apartments: a large number of buildings connected by an articulated plan, that aims to become a new attraction of the city. The aggregate of buildings thus becomes a composition of free and asymmetric forms characterized by an organic architecture that gives the whole unique chiaroscuro effects and allows it to blend into the surrounding, mainly flat territory and move its skyline through a new system of levels. The facing of the architectural volume evokes a textile design that enhances the image of the area; rather than contradicting its essence, it accentuates its identity.
.pag 756-765

**Location** Changsha
**Project** Changsha Mei xi lake international culture and art center Competition
**Client** Competition
**Plan** 2011
**Plot area** 200000 mq
**Built area** 320000 mq

## BODY'S GYM
Florence, Italy, 1998-2000

The property in question is hidden in the dense texture of an urban fabric characterized by the formal uniformity typical of the intensive building of the 1960's. The redevelopment program includes the transformation of an artisanal warehouse, located among residential buildings, into a place for recreation, fitness, and health treatments. The project plays the role of social catalyst of today's modern wellness temples, called on to animate the context of which they are part of.
Having the need to set itself into the continuity of the existing building led to the search for the architectural quality of "difference", through emphasizing the identity of the façade. The entrance wall, made of offset mounted double steel mesh, was actually conceived as a semitransparent diaphragm, that allows the building to continually change its image according to the hour of the day: an opaque screen in the daytime, iridescent veil at night. The interior arrangement finds its formal and functional reference in the Corten steel calendered staircase, which evolves the plan's circularity like a propeller, wrapping around itself and supporting the projecting stair treads.
**pag 768-773**

**location** Florence, Italy
**project** Fitness centre
**client** Body's Gym S.r.l.
**plan** 1998
**construction** 1999-2000
**cost** € 700,000
**built area** 1,000 sq.m
**volume** 4,000 cu.m
**contractor** Impresa Edile Turtora Cannizzaro

## TORNABUONI ARTE GALLERY
Portofino, Italy, 1999

The project involves a small warehouse located in exclusive via Roma, next to the oratory of Our Lady of the Assumption. The "reuse" plan for the building as an art gallery required a redevelopment and restoration that was conservative relative to the technical construction aspect as well as to the public image of the building itself. Externally this latter theme reintegrates it with its prestigious urban surroundings: the new slate covered timber roof continues into the front portico, replacing the dilapidated sheet roof, while the facade plaster is linked chromatically to the adjacent oratory. Inside the slightness of the simple narrow and long rectangular space provides the opportunity to characterize the environment through the surfaces that delimit it. Corten steel panels are completely cover the interior space and make up the vertical support for the paintings arranged along the side walls. One of them is equipped with a mechanical system capable of making these panels mobile and interchangeable, in order to multiply the display area and simultaneously restore an image of the dynamic spatial configuration, able to surprise the visitor with continuing variation.
**pag 774-779**

**location** Portofino, Genoa, Italy
**project** Commercial, exhibition
**client** Tornabuoni Arte S.r.l.
**plan** 1999
**construction** 1999
**cost** € 350,000
**built area** 30 sq.m
**contractor** Fratelli Giani S.r.l.

# 梅溪湖(MADAME DAI)文化艺术中心

长沙，中国，2011年

本项目位于总体规划的东北方位，且该总体规划是经由纽约的KPF公司构建的，旨在实现梅溪湖畔区域的新型城市化，而梅溪湖为坐落在大河西地区的一个超过40公顷的大型人工湖。作为将被打造成长沙地区可持续发展进程中的一个试点项目，该新文化中心与城镇规划的有机设计相得益彰并符合本项目的理念，而其中的虑及方面包括可持续性、人为活动和自然景观的一体化，并参谋了地域的文化特征。

该综合设施包含了诸多不同功能设施：一所剧院、一处画廊和一间礼堂 – 融合在一个单一的地域之中– 以及一处购物中心、一家酒店、许多办公室和公寓：而诸多建筑物通过一套链接式规划被连为一体，旨在打造成为本城市的一个新的魅力点。

本栋总体建筑将自此主打自由和非对称形式的牌，以其有机架构的特点展现全方位及独特的明暗对比效果，并致力于与周边多显平坦的环境相互呼应，且通过一套崭新的层次系统沿地平线拓展开来。建筑物的饰面采用为织物设计，以便增强区域的视觉效果；而不是与其本质相互矛盾，也即突出了身份。

页号756-765

地点　　长沙
项目　　长沙梅溪湖国际文化艺术中心竞赛
客户　　竞赛
规划　　2011年
占地面积　200000平方米
建筑面积　320000平方米

# BODY健身馆

佛罗伦萨，意大利，1998年-2000年

项目主体属性被隐藏在城市密集的纹理当中，这是一种具有20世纪60年代密集型建筑物典型的正规而统一的特色。这个新建计划包括了对工匠仓库的改造，此仓库位于居民建筑物之中，改造将把它转化成一个娱乐、健身和养身理疗场所。这个项目在当今的现代化健康模板当中起着社会催化剂的作用，呼吁着自己也是使社会环境焕发生命力的一部分。

由于建筑需要设置在连续的现有墙体中，导致我们要通过强调建筑立面的特征来寻求表现"不同之处"的建筑品质。入口墙壁外挂双层钢丝网，制造出半透明隔膜的效果，使得这座建筑物能够根据一天当中随着时间的推移连续地改变自身形象：在白天是一个不透明的帷幕，在夜晚是色彩斑斓的面纱。内饰布置可以在考登钢楼梯上找到其形式上和功能上的参照物，这个楼梯就像一个螺旋桨一样沿环形升高，自我缠绕着并支撑着凸出的楼梯台阶。

页号768-773

地点　　佛罗伦萨，意大利
项目　　健身中心
客户　　Body's Gym S.r.l.
规划　　1998年
建设　　1999年-2000年
成本　　700,000欧元
建筑面积　1,000平方米
体量　　4,000立方米
承包商　Impresa Edile Turtora Cannizzaro

# TORNABUONI ARTE画廊

Portofino，意大利，1999年

项目位于Via Roma上的一个小仓库，紧挨着Nostra Signora dell'Assunta小礼拜堂之后。将该建筑作为艺术画廊重新使用，要求在技术构造方面和建筑物公共形象上进行重建和修复。从外观上看，建筑需要和尊耀的周边环境相融合，木屋顶的新板岩瓦延续至前部的柱廊，取代了年久失修的顶板，同时建筑正面的灰泥也与邻近的小礼拜堂从色调上形成了关联。在这个简单的狭窄而细长的长方体空间的纤细空间界限内，提供了通过界定此空间的表面来描述环境特色的机会。考登钢壁板完全覆盖了内部空间并构成了沿侧墙布置的油画作品的垂直支撑。其中一个支柱配备了能够让这些壁板移动和互换的机械系统，可以让展示区域倍增，并同时恢复了动态空间配置的形象，能够以连续不断的变化给参观者带来惊喜。

页号774-779

地点　　Portofino，热那亚，意大利
项目　　商业，展览
客户　　Tornabuoni Arte S.r.l.
规划　　1999年
建设　　1999年
成本　　350,000欧元
建筑面积　30平方米
承包商　Fratelli Giani S.r.l.

## GRANITIFIANDRE SHOWROOM
**Castellarano, Italy, 2001-2002**

The new megastore is found located in a vast area in the immediate vicinity of the Castellarano headquarters, on the Emilian plain. The plan combines the normal activities of the sales and marketing functional units such as library, and design workshop, at the end to communicate to visitors the company's philosophy. The project concept is based on an analogy with the spatiality of the quarry and the open process of production and invention of the materials derived from it. Spaces and their functions are crystallized in five stereometric clear glass volumes that stand out like iridescent plates on a gravelly layer on top of which the connecting pathways are articulated. To the naturalization of the environment there corresponds the sophisticated technology of the sound and lighting system, which enables the activation of functions and lighting to the visitor pathways, which guide them through a sort of narrative through the various working phases, described by a series of backlit screened prints. The chromatic variety of product samples, set along the wall down the main access path, is enhanced by the light which is diffused from the "Michelangelo Statuary" plate utilized for the coverings.
**pag 780-785**

**location** Castellarano, Reggio Emilia, Italy
**project** Showroom
**client** GranitiFiandre S.p.A.
**structures** Studio Tecnico Cuoghi
**plan** 2001
**construction** 2001

## BALUARDO SAN COLOMBANO
**Lucca, Italy, 2002-2003**

The walls of Lucca, since the 1.500's the architectural boundary between the ancient and new city, are experienced by residents as a public space and also an exhibition center and place where cultural events are held, with noted intellectual circles like the historical Caffè delle Mura. From here the competition for the transformation into a café-restaurant at the rampart of San Colombano, and the idea, expressed in the project proposal, is to make the life of the locale the protagonist in the life of the city. The transparency makes customers' visits visible, and interprets the relationship between the interior "scene" and the vivid urban environment, while the oak floor boards allow, as in a theater stage, the recording also acoustically of customers' presence. The same essence is used for the roof and the long bar counter, an organic layering of splinters that evoke Ceroli's sculpture. The partitions between the reception and service spaces are made from treated iron, like sliding fixtures. The use of the materials, like the study of every other element, from furniture to lighting, helps to characterize the space as a "total design" of integration, respecting and harmonizing with the place's history. The project is conceived so as to be completely reversible and detachable, bringing the rampart back one day to its original configuration.
**pag 786-795**

**location** Lucca, Italy
**project** Restaurant and lounge bar
**client** Carmafrigor S.r.l.
**systems** P.I. Luca Pollastrini
**plan** 2002-2003
**construction** 2003
**cost** € 600,000
**built area** 200 sq.m
**contractor** Michele Bianchi S.r.l.

## TORNABUONI ARTE GALLERY
**Venice, Italy, 2004**

The project involves a building in Campo San Maurizio, Piazza San Marco area. The restoration plan of a store-front commercial space envisaged its transformation into the Tornabuoni Arte contemporary art gallery, after those already active in Florence, Milan, Portofino, and Forte dei Marmi. The project is an opportunity to experiment with the theme of the small scale exhibition space. The paucity of available surfaces suggested a concept based on the perception of space and the relationship between container / contents. The existing structure, the result of an earlier restoration of the historical building, is subject to a further construction intervention aimed at the total elimination of the vertical supports: this allows the creation of an organic and winding indirect route. The white resin wraps indiscriminately around the horizontal and vertical surfaces sublimating the structure in a sculptural unity from which all the constituent and accessory spatial elements take their form, including the seating and horizontal surfaces. The lights arranged between the "craters" in the ceiling are the only "detail" in a spatially monochromatic continuum, smooth and devoid of references to geometric perspective, except for the works of art on display.
**pag 796-801**

**location** Venice, Italy
**project** Commercial, exhibition
**client** Tornabuoni Arte S.r.l.
**structures** Favero&Milan Ingegneria
**plan** 2004
**construction** 2004
**cost** € 700,000
**built area** 60 sq.m
**contractor** Friulan

## GRANITIFIANDRE陈列室
**Castellarano，意大利，2001年-2002年**

这座新的特大型商店紧挨位于Emilian平原上的Castellarano总部的一块巨大区域内。这个计划包括了销售和营销职能部门的正常活动场所，诸如图书室、设计工作室，并最终向来访者传达公司的经营理念。这个项目的设计概念以对采石场空间属性的模仿以及开放性生产过程和材料发明工作为基础。各个空间及其功能都在5个立体透明玻璃体量内具体表示出来，这些体量就像铺在碎石层上的珠光玻璃釉彩板一样，碎石层上连接着用于连通各处的小径。为了实现环境的自然化，对应采用了技术复杂的声响和照明系统，使得各项功能运作并可以对参观者路径进行照明，而这些路径将参观者引导至一种通过各个工作阶段展开的叙述过程，并由一系列背光照明的印刷品进行文字描述。产品样品的色彩变化度沿着主要进入通道的墙壁设置，并由当作吊顶板的"Michelangelo Statuary"板所漫射而来的照明光进行增强。
**页号780-785**

| | |
|---|---|
| **地点** | Castellarano，Reggio Emilia，意大利 |
| **项目** | 陈列室 |
| **客户** | GranitiFiandre S.p.A. |
| **结构** | Studio Tecnico Cuoghi |
| **规划** | 2001年 |
| **建设** | 2001年 |

## BALUARDO SAN COLOMBANO
**卢卡，意大利，2002年-2003年**

卢卡城墙自1500年起就是古城与新城之间的建筑边界，曾经被居民们当作公共空间来使用，也当作过一个举办各种文化活动的展览中心和场所，拥有类似于历史性的Caffè delle Mura一样的知识分子聚集点。以此为出发点，这次设计竞赛针对的是其改造成一个位于San Colombano城墙上的咖啡馆－餐馆，而在项目建议书中所表达的创意就是，让这个城市生活中主角故事发生的地点变得鲜活起来。透明材质可让人们看到顾客的到来，并解释了内部"场景"与生动城市环境之间的关系，而橡木地板可以作为一个剧院舞台，而录音也从声音上宣示了顾客的存在。同样的特质也应用于屋顶和长酒吧柜台，一层有机碎片覆盖层唤醒人们对于Ceroli雕塑的记忆。接待区与服务空间之间的分隔墙采用经处理的铁质材料来制作，诸如滑道固定装置。对材质的运用就像每一个其它元素一样进行了研究，从家具到照明，都有助于描述这个空间作为集成"总体设计"的特色，对应于这个场所的历史并协调一致。项目构思为完全可逆转和可拆卸的方式，只需一天就可将城墙恢复至其原始的形态。
**页号786-7951**

| | |
|---|---|
| **地点** | 卢卡，意大利 |
| **项目** | 餐馆和雅座酒吧 |
| **客户** | Carmafrigor S.r.l. |
| **系统** | P.I. Luca Pollastrini |
| **规划** | 2002年-2003年 |
| **建设** | 2003年 |
| **成本** | 600,000欧元 |
| **建筑面积** | 200平方米 |
| **承包商** | Michele Bianchi S.r.l. |

## TORNABUONI ARTE画廊
**威尼斯，意大利，2004年**

此项目涉及一座位于 Piazza San Marco区Campo San Maurizio的建筑物。对一座商店的后部的重建，将其改造成为Tornabuoni Arte现代艺术画廊，继该画廊在佛罗伦萨，米兰, Portofino以及Forte dei Marmi活跃经营后，这个项目是一个试验小尺度展览空间主题的机会。可供使用表面的不足让我们构思出一种以对空间以及包容物/内容物之间关系的感受基础上的设计概念。现有购结构是对这座历史建筑早期改造的结果，将进行进一步建设以达到完全消除垂直支撑的目的：这样允许形成一个有机而蜿蜒的间接路线。白色树脂无差别地包裹着水平和垂直表面，使这幢构筑物升华到一种雕塑般的统一，并由此赋予了所有组成和附属空间元素的造型形式，包括座位和水平表面。照明灯光布置在天花板内的"陨石坑"内，它也是唯一在空间上具有单色调连续性的"细节"，除了所展示的艺术作品以外，它们表面光滑并缺乏与几何透视关系的参照。
**页号796-801**

| | |
|---|---|
| **地点** | 威尼斯，意大利 |
| **项目** | 商业，展览 |
| **客户** | Tornabuoni Arte S.r.l. |
| **结构** | Favero&Milan Ingegneria |
| **规划** | 2004年 |
| **建设** | 2004年 |
| **成本** | 700,000欧元 |
| **建筑面积** | 60平方米 |
| **承包商** | Friulan |

## LONELY LIVING
### Venice, Italy, 2002

The exhibition "Lonely Living. The architecture of the primary space" was held In the sphere of the VIII Venice Architecture Biennial. Eighteen 1:1 scale building models that the architects selected have been drawn up for as many hypothetical clients. The project, dedicated to the artist Franco Ionda, is based on an idea of material subtraction: a 16 mc wooden cube, was excavated until the identification of a Laugerian "primitive hut", which houses the few but essential vital functions of a life devoted to art. The perception of the space is beyond the categories related to dimension and use: the bed/shelter for sleeping and reflecting is a large tub filled with linseed oil whose smell contributes to the deep immersion of the senses in a lived work of art, which translates living as a pure affirmation of being. The building, between construction and grotto, between ravine and elementary cell, is made up by the summation of overlapping layers of wood: the assembly is designed as an Etruscan pseudo-vault composed of parallel planes, programmatically devoid of technology or constructive refinement.
**pag 804-809**

**location** Giardini di Castello, Venice, Italy
**project** Exhibition, set
**client** Aid'A Agenzia Italiana di Architettura
**structures** A&I progetti - Niccolò De Robertis
**systems** Martini Illuminazione - Alberto Mantovani
**plan** 2002
**construction** 2002
**built area** 9,00 sq.m
**volume** 270 cu.m
**contractor** Fima Cosma Silos

## ENZIMI
### Rome, Italy, 2003

The Enzimi event has supported and promoted, ever since mayor Veltroni meant to further creativity and multidisciplinary and multiethnic cultural comparison in Rome. The 2003 edition involves Piazza Vittorio Emanuele and the architectural work of the Esquilino; the program involves the inclusion of informative and functional elements, areas for dancing, entertainment, music, photography, games, and refreshments. The project is based on a staging capable of physically materializing the event through the scenic representation of a giant souk, of an Arab city, proposed as an alternative urban configuration to the existing one. This landscape of new elements is consolidated on an enormous elevated aluminum footboard which redefines the square and arranges the paths, identifying the exhibition places that are supported on a sort of "new floor". This thin layer, hardly a step, is inhabited by volumes obtained from the various aggregation of cubic elements, made up of a modular metal framework closed by translucent polycarbonated surfaces. These rise in every direction appearing by day as opaline boxes and by night like luminous architectural beacons. The diagram rationalizes and synthesizes the spaces making it easy to put the messages and information related to the event in order.
**pag 810-815**

**location** Rome, Italy
**project** Exhibition, set
**client** Zone Attive – Comune di Roma
**plan** 2003
**construction** 2003
**cost** € 150,000
**aluminum area** 1,600 sq.m
**polycarbonate covering area** 1,884 sq.m
**volume** 2,240 cu.m
**contractor** Nolostand

## VINAR
### Florence, Italy, 2005

The staging for the Vinar exhibition occupies the striking setting of Stazione Leopolda in Florence. The event is centered on the interaction between enologic culture, art, and architecture. The intention is to recreate atmospheres, smells, and colors of the wine world, between nature and artifice. The staging is arranged in six thematic sections distributed longitudinally through the building. The entrance is dominated by green light and by a huge chandelier, made up of 1,800 bottles, suspended over the bookshop. The tasting area, bounded by a very long wing of industrial wooden pallets, is developed laterally and in parallel with a 100 meter long table fronted by 25 bright vats and 300 oak barrels, against the background of a "tent" made of baguettes of bread, which concludes the path. In the space called "Alcatraz", above a layer of earth arranged on the floor, a real lawn, covered by a carpet of leaves, where it is possible to see, like in the middle of a forest, Dimitri Kozaris' video installation on wine and the history of cinema, followed by, at the end of the central nave, Paolo Fiumi's multisensory installation. This area also hosts "minimal diatribes", meetings on the theme with architects, artists, journalists and writers. Along the right side of the ex-station Leopolda the walls are used as a projection place for the section called "the author's cellars".
**pag 816-823**

**location** Stazione Leopolda, Florence, Italy
**project** Set
**client** Federico Motta Editore and Pitti Immagine
**plan** 2005
**construction** 2005
**cost** € 300,000
**built area** 600 sq.m
**contractor** Machina S.r.l.

# LONELY LIVING

威尼斯，意大利，2002年

这个展览称为"孤独生活"。主要空间的建筑设计（Lonely Living. The architecture of the primary space）"在第8届威尼斯双年展期间内举办。建筑师选择的8个1:1比例的建筑模型吸引了许多潜在客户。这个项目是献给艺术家Franco Ionda的，并以材料减法的创意为基础：1个16立方米的木制立方体，被挖空至形成Laugerian"原始小屋"的标识为止，其中容纳着几项对于一个完全献身于艺术的生活十分重要的功能。对这个空间的感受超越了与尺度和用途相关的类别：用于睡眠和沉思的床，遮蔽物是一个注满了亚麻籽油的大浴盆，它的气味有助于使人深深沉浸在对生活艺术的感受中，这个作品将生存转化成纯粹的对存在的感知。这座建筑物与天然洞穴之间，位于建筑遗迹与住宅基本单人房之间，由相互重叠木头层汇集而成：这个组件设计成一种伊特鲁里亚伪拱顶的形式，由多个平行平面构成，有规划性地避免了技术或建筑上的精细化。

页号804-809

地点　Giardini di Castello，威尼斯，意大利
项目　展览，装置
客户　Aid'A Agenzia Italiana di Architettura
结构　aei progetti - Niccolò De Robertis
系统　Martini Illuminazione - Alberto Mantovani
规划　2002年
建设　2002年
建筑面积　900平方米
体量　270立方米
承包商　Fima Cosma Silos

# ENZIMI

罗马，意大利，2003年

Enzimi活动自从市长Veltroni倡导以来，一直促进着罗马进一步发展创新和多学科和多种族的文化比较。2003年版本涉在Vittorio Emanuele广场的Esquilino的建筑设计作品；这个计划涉及到信息性和功能性元素包含供舞蹈、娱乐、音乐、摄像、游戏和休闲娱乐的区域。本项目以能够从实体上体现此次活动的展示为基础，通过再现阿拉伯城市的大型鲁天市场，提议将其作为现有城市结构的代替方案。新元素的景观统一到数量众多的升高铝板踏脚板上，这些踏脚板重新界定了这个广场并安排了路径，标识出在一种"新地板"上得到支撑的展览场所。这个薄薄的一层，几乎不到一个台阶，容纳不同大小的正方体构件，由金属框架和全封闭的半透明的聚碳酸酯构成。这些升高部分在白天从各个方向上显示为乳白色的盒子，而在夜晚就像会发光的建筑灯柱一样。这个设计方案让这些空间合理化，合成化，让与活动相关的信息与消息更易分门归类。

页号810-815

地点　罗马，意大利
项目　展览，装置
客户　Zone Attive – Comune di Roma
规划　2003年
建设　2003年
成本　15,000欧元
铝制品面积　1,600平方米
覆盖聚碳酸酯的面积　1,884平方米
体量　2,240立方米
承包商　Nolostand

# VINAR

佛罗伦萨，意大利，2005年

Vinar展览的展出营造了佛罗伦萨前火车站Stazione Leopolda的奇特布景。此项活动中心点是葡萄酒酿造文化、艺术和建筑设计的互动。其宗旨是重现葡萄酒王国处于自然和人造物之间的氛围、气味和色彩。展出安排为6个主题部分，在整个建筑物内沿纵向分布。入口处被绿色灯光和一个巨型枝形吊灯所主宰，这个吊灯由1800个酒瓶制成，悬吊在书店的上方。品酒区由一个非常长的工业木板箱所界定，它沿横向布置并与一条100米长的桌子平行，正面布置了25个光洁的大桶和300个橡木酒桶，并以用法国长棍面包制成的"帐篷"为背景，此帐篷作为路径的终点。在称为"Alcatraz"的空间内，在地板上铺好的泥土上，布置一块真草坪，用叶子组成的地毯进行覆盖，就像可以在森林中间看到的一样，在Dimitri Kozaris的有关葡萄酒和电影院的视觉装置之后，在中央大厅的端头，布置着Paolo Fiumi的多感官装置艺术。这个区域还容纳着"minimal diatribes"，一个由建筑师、艺术家、记者和作家共同举行的会议方式。沿前Leopolda车站的右侧的墙壁被用作一段称为"作家的密室"展览部分的投影场所。

页号816-823

地点　Stazione Leopolda，佛罗伦萨，意大利
项目　装置
客户　Federico Motta Editore and Pitti Immagine
规划　2005年
建设　2005年
成本　300,000欧元
建筑面积　600平方米
承包商　Machina S.r.l.

## LABORATORIO ITALIA
**Rome, Italy, 2006**

The preparation made in Rome at the Sala Clementina of the former juvenile prison of San Michele concludes and relaunches a research path started about a year ago by d'Architettura magazine. The essential contents of two previous exhibitions – The Parma Architecture Exhibition; L(es) Etranger(es) in Brescia, – have marked the gradual defining of the initiative, aimed at jointly revealing the objective of the investigation: opening a breach onto the state of architecture in Italy, regardless of the nationality of the designers and from the centrality of the descriptive appearance of the works, shifting attention towards the complexity of the thematic aspects that intercept the architectural culture of our country. The research material collected by a team of young architects and organized into a "choral" narrative structure, which places next to the project images a series of interviews and reflections that "report" from inside, the current condition of the architect in Italy: the different figures collected within a single definition; the relationship with education, models, languages; the problems related to the professional accounting year, etc. The expositive-narrative path is defined by the presence of a red plasterboard central structure, whose broken configuration moves the geometric rigor of the room. The longitudinal development of the latter is emphasized by the support-walls – in black plasterboard – that trace the perimeter.
**pag 824-827**

**location** Rome, Italy
**project** Exhibition, set
**client** Aid'A Agenzia Italiana d'Architettura, DARC Direzione Generale per l'Architettura e l'Arte Contemporanee
**plan** 2006
**construction** 2006

## ANNALI DELL'ARCHITETTURA 2007
**Naples, Italy, 2007**

The staging for the 2007 edition of the Annals dell'Architettura e delle Città, entitled "Mediterranean Nomadism", is placed in the spaces of the Palazzo Reale in Naples, attempting to physically underline the presence and importance of the sea, that scenario of travel between geographically and culturally distant countries, albeit connected by a common destiny. The project interprets the theme through an ordered series of port containers that invade the porticos on the Piazza del Plebiscito, the courtyard and the ambulatory, imposing themselves also on the urban scenario like "thresholds" in correspondence with the entrances of the two squares. Associated with the concept of commerce these iron yellow boxes transform the Palazzo Reale into a sort of giant ship transporting containers loaded with information, and culture. In particular, the section dedicated to the architecture of the Giant Ships occupies a series of overlapping containers, inset into the archways that design the sides of the court. These artificial spaces, utilized as actual "cameras obscura" allow the projection of filmed compositions that tell the story of the great Italian shipbuilding and cruising tradition. Other containers placed at the court's center are used to narrate the events of past emigration and of the tragic clandestine immigration landing on the Italian coast every year.
**pag 828-835**

**location** Palazzo Reale, Naples, Italy
**project** Exhibition, set
**client** Fondazione Annali dell'Architettura e delle Città, Napoli
**plan** 2007
**construction** 2007
**contractor** Domino S.r.l.

## 1968/2008 QUARANT'ANNI DI DESIGN
**Milan, Italy, 2008**

The preparation of the exhibition forms part of the events surrounding the Salone del Mobile festival in Milan. The concept is the result of research work conducted by the journal Area for the publication of the special edition "97+", dedicated to the forty years of design between 1968 and 2008. The juxtaposition of objects from different eras becomes the pretext for an interactive reflection on the design project: the object as a document of the evolution of modern society in the last 40 years, with the relative changes in cultural and stylistic patterns of life. Inside the exhibition space, the Collina Hall at the headquarters of Il Sole 24 ORE, a new volume defined by a textile covering measured by a sequence of bands like the pages of a magazine has been constructed: they are raised off the ground to become, by conjuring pages of images, the three-dimensional support of real objects. The exhibit scheme is arranged in pairs of objects, similar by type and function, but distinct by production eras, like being in front of a "time mirror"; the substantial caption text contribution accompanies the visitor by highlighting significant similarities and differences. On the back wall the flow of video images mixed dialogue among the greatest exponents of the Italian design tradition with cinematographic excerpts drawn from the repertory of the last 40 years of the daily "consumption" of design.
**pag 836-843**

**location** Il Sole 24 ORE headquarters, Milan, Italy
**project** Exhibition, set
**client** Il Sole 24 ORE Business Media
**plan** 2008
**construction** 2008
**contractor** Machina S.r.l.

## 意大利实验室
罗马，意大利，**2006年**

在罗马的San Michele前青少年监狱Sala Clementina所做的准备工作结束并重新启动了一条一年以前由d'Architettura杂志启动的研究之路。两项之前举办的展览 – 帕尔马建筑设计展；L(es) Etranger(es) in Brescia的精华内容标志着此项倡议的逐渐界定，目标放在联合此项调查：为意大利建筑设计的状态开创一次突破，无论设计师的国籍以及作品描述外观的集中性，将注意力转向主题方面的复杂性，这种复杂性拦截着意大利本国建筑设计文化的发展。研究材料由一个年轻建筑师团队来搜集并组织成"唱诗班"式的叙述结构，在项目图像之后放置一系列的访谈和沉思，从内部"报告"意大利建筑设计的当前状况：在同一定义内所搜集的不同数字；与教育、模型、语言之间的关系；与专业会计年度相关的问题与德行。而红色石膏板中央结构的存在界定了这种解释叙述的路径，这个结构的破裂性形态改变了这个房间的几何线条严谨性。后者的纵向延伸被支撑墙壁所强调墙壁采用了黑色石膏板并沿周边布置。
**页号824-827**

**地点** 罗马，意大利
**项目** 展览，装置
**客户** Aid'A Agenzia Italiana d'Architettura, DARC Direzione Generale per l'Architettura e l'Arte Contemporanee
**规划** 2006年
**建设** 2006年

## 建筑编年史 2007
那不勒斯，意大利，**2007年**

2007年的建筑与城市编年史展出，标题为"地中海游牧民"，安排在那不勒斯Palazzo Reale的空间进行，试图从实体上突出海洋的存在和重要性，通过地理上和文化上都距离非常遥远的国度之间的旅行场景，展出它们共同的命运。这个项目通过在Plebiscito广场的柱廊、庭院和回廊的有序排列一连串港口集装箱来解释这个主题，还将它们强加于对应于两个广场入口的诸如"门槛"这样的城市场景。与商业概念相关联的这些铁黄色的箱子将Palazzo Reale改造成为一种巨型集装箱运输船，船上装载着信息和文化。特别是，专门用于这只巨船建筑设计的部分占据了一串相互重叠的集装箱，并插入到构成了庭院侧面部分的拱道内。这些人造空间被作为真正的"相机暗箱"使用，实现了胶片组成部分的投影，而胶片部分讲述了伟大的意大利造船和巡航传统。其它放在庭院中心的集装箱用于叙述以前移民的事件以及每年登陆意大利海岸的悲剧性偷渡事件。
**页号828-835**

**地点** Palazzo Reale，那不勒斯，意大利
**项目** 展览，装置
**客户** Fondazione Annali dell'Architettura e delle Città, Napoli
**规划** 2007年
**建设** 2007年
**承包商** Domino S.r.l.

## 1968/2008 四十年设计
米兰，意大利，**2008年**

这项展览是围绕米兰Salone del Mobile节的活动的一部分。设计概念是由Area杂志为了出版其专刊"97+"而展开的研究工作的成果，这份专刊专门叙述了1968年至2008年这40年间的设计。将来自不同时代的物品并列已成为这个设计项目的互动性沉思的声明：其目的是成为记载现代社会在过去40年演变以及文化和生活风格模式上变化的记录档案。在这个展览空间之内，在Il Sole 24 ORE总部的Collina厅，构建了一个新体量，这个体量由一个纺织材料覆盖物来界定，一系列类似于杂志页面的条带有规律地布置。这些部分升高离开地面，通过魔术般地呈现影像页面，成为了真实物品的三维支持。展览主题以展品对比的方式安排，其功能相似，但生产年代截然不同，就像站在"时光镜"前面一样；丰富的大写字母文稿通过强调显著的类似性和差异性为参观者提供了解析。在后墙上，视频影像的流动混合着意大利设计传统最伟大倡导者的对话，并且展示了自过去40年日常"消耗品"设计的保留剧目中抽取的电影摘录。
**页号836-843**

**地点** Il Sole 24 ORE总部，米兰，意大利
**项目** 展览，装置
**客户** Il Sole 24 ORE Business Media
**规划** 2008年
**建设** 2008年
**承包商** Machina S.r.l.

# ANNALI DELL'ARCHITETTURA 2008
**Naples, Italy, 2008**

The 2008 edition of the Annals expresses the programmatic intent of comparing the culture of architectural design with the issues of land management that need immediate responses. The provocative title "Mediterranean Emergency" refers to the deplorable situation that has seen Naples in the worldwide headlines, and by extension, to all those critical situations that threaten all of the "Mediterranean" areas, identifying areas of recovery as well as forecasts of possible eco-compatible development. This involves the elaboration of reference models that make available innovative solutions in regards to both systems and landscape redevelopment. The exhibition is divided into two thematic sections. The first occupies the Court of Honor of the Palazzo Reale with a 400 sq.m elevated platform on which the mapping of the Campania region is reproduced, indicating the centrality given by the event to the status of the management of this specific geographic area both in terms of the detection of situations and of proposed redevelopment. From here the theme of environmental emergency extends to national and international case studies: a one-meter wide ribbon floor, conceived as a sequence of text and images, accompanies visitors toward the ambulatory where the exhibition continues with the section dedicated to examples of landfill thermo-reevaluation and redevelopment systems.
**pag 844-851**

**location** Palazzo Reale, Naples, Italy
**project** Exhibition, set
**client** Fondazione Annali dell'Architettura e delle Città, Napoli
**plan** 2008
**construction** 2008

# THE TABLE OF ARCHITECTURE
**Padua, Italy, 2008**

The completion of the work is located in the "Barbara Cappochin" area of the Architecture Biennial, the event held in Padua in the Palazzo della Ragione. The project concerns the exhibition tied to the namesake International Prize, to which is connected a section dedicated to the architects e the projects completed in Padua's urban area, organized by the provincial order of architects. The Table interprets exhibition staging as an architectural "communication" strategy. The idea is to bring the project out from the inside of one of the many city palaces to make it "come down to the street", intercepting the urban itineraries, where the inhabitants can simply observe them passing. The table – setup in the central and crowded Piazza Cavour – underlines the desire to construct a symbolic place of cohabitation but also of dialogue and debate. The object, purposely spartan, is made up of a succession of two prefabricated reinforced concrete elements (used for the construction of industrial building or parking structure roofing), 12 m in length, merged to form a single 24 meter long "Banquet" display, made with iron oxide enriched cement. The images, taken from the catalog pages, have thus been resined onto the table surface that makes up the actual expositive area of this section of the exhibition. The success of this initiative, attested to by the participation of the guests, has made it one of the icons of the biennial exhibition.
**pag 852-855**

**location** Padua, Italy
**project** Exhibition, set
**client** Fondazione Barbara Cappochin
**plan** 2008
**construction** 2008

## 2008建筑年鉴

**那不勒斯，意大利，2008年**

## 建筑表

**帕多瓦，意大利，2008年**

Annals展览的2008年版表达了这样一种纲领性的意图，想要将建筑设计文化和土地管理急需应对地问题进行对照。展览的倡导性标题"地中海紧急事件"指的是导致那不勒斯出现在全球新闻头条的可叹状况，通过拓展性思考，让我们考虑到那些威胁着所有"地中海"地区的危急状况，确立那些要恢复的地区以及预测可能的与生态系统相容的开发行为。这涉及到能够从系统和景观重新开发方面提供创新解决方案的借鉴模型。本展览分为两个主题部分。第一个为在皇宫地"荣耀庭院"占据400平方米地架高平台，平台上复现了坎帕尼亚区的地图，指出了此展览的主题是有关这个具体地理区域状态的管理方面，这种管理既涉及到对状况的检查，又涉及到所设想地再度开发。在此，环境紧急事件的主题延伸到了国家和国际性的案例研究：1米宽的缎带地板，构思为一系列的文本和图像，引导着参观者走向回廊，这里继续展出本展览关于垃圾填埋场热量重新评估和重新开发系统的示例。

**页号844-851**

**地点** Palazzo Reale，那不勒斯，意大利
**项目** 展览，装置
**客户** Fondazione Annali dell'Architettura e delle Città, Napoli
**规划** 2008年
**建设** 2008年

此项作品的完成版位于在Palazzo della Ragione宫内，由帕多瓦省举办的"建筑设计双年展"的"Barbara Cappochin"展区。此项目是为同名国际奖项地展览策划地，该奖项设有一单元，由省级建筑师协会评选在帕多瓦地区完成地建筑项目及其建筑师。这个建筑表，展示了作为展览作为"交流"策略，将展览安排在市民随性徜徉就能观摩展览地马路上。这个建筑表 – 安放在市中心Cavour广场上– 强调了构建共居符号性场所对话和辩论的愿望。这个对象特意做得十分简朴，由一系列的两个预制钢筋混凝土构件（用于建造工业建筑或停车场构筑物屋顶）组成，长度为12米，合并起来形成一个单条24米长的"宴会（Banquet）"展览，并采用氧化铁水泥制作。因此，从目录页面上取得的图片已经对表格表面进行了树脂涂敷，而这些表格的表面就构成了这部分展览的实际展览区域。这个倡议的成功受到了来宾参与的证明，并成为本次双年展的代表作品之一。

**页号852-855**

# LANDSCAPE

景观设计

# ACOUSTIC BARR

## Impruneta, Florence, Italy, 2000-2007

地点　意大利佛罗伦萨Impruneta镇
项目　吸音和反射声波屏障
客户　Autostrade per l'Italia S.p.A.
结构　CIR-Ambiente
规划　2000年
施工　2005年-2007年
成本　6500000欧元
建筑面积　8,300,000平方米
赤陶所覆盖表面积　20000平方米
承包商　CIR-ambiente

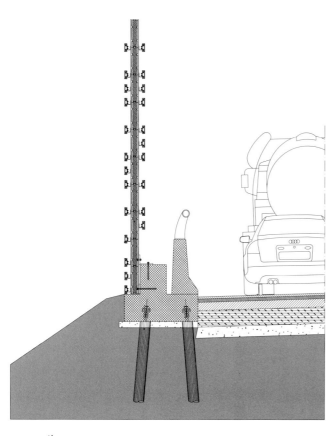

cross section
横剖面

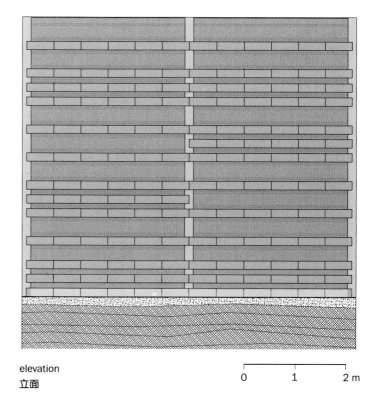

elevation
立面

0    1    2 m

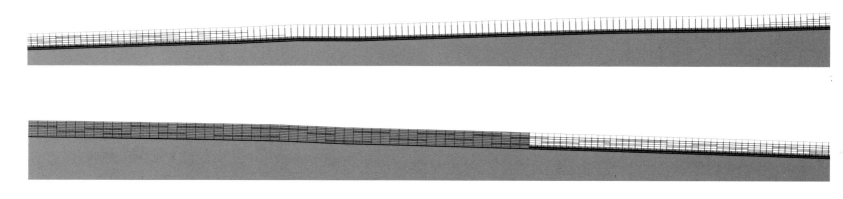

highway stretch Pozzolatico – external side highway
波佐拉提科公路延伸——外部公路

0    10    20 m

# VIA TIRRENO REDEVELOPMENT

## Potenza, Italy, 2000-2010

地点　意大利，Potenza
项目　城市更新
客户　Potenza市政府
结构　Favero&Milan Ingegneria
系统　Favero&Milan Ingegneria
规划　2000年
建设　在建 2006年-2010年
成本　2066737欧元
建筑面积　3248平方米
体量　6800立方米
承包商　Giovanni Basentini Lavori S.r.l. C.S.T. impianti S.r.l.

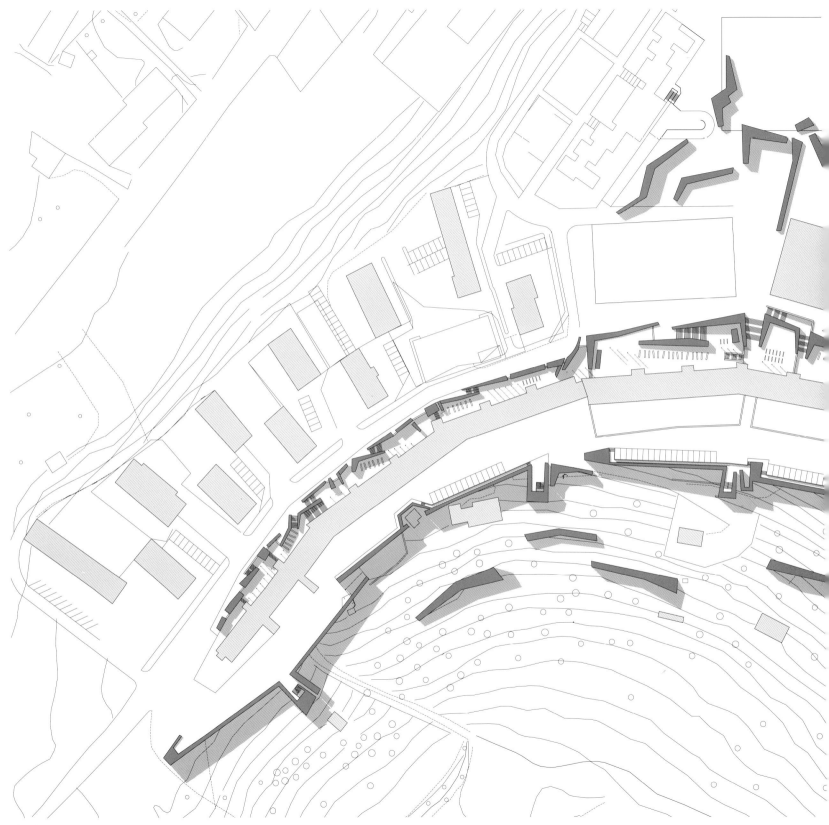

site plan
位置图

0    20    50 m

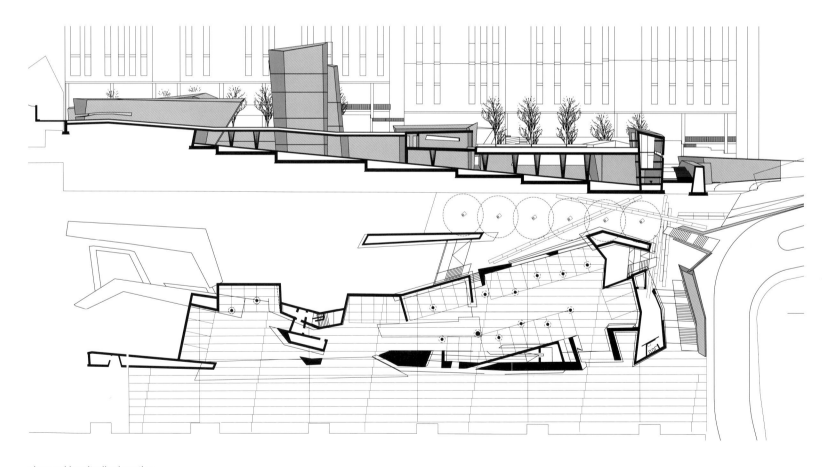

plan and longitudinal section
平面和纵剖面

0    10    20m

# ANTINORI WINERY

## San Casciano Val di Pesa, Florence, Italy

## 2004 - under construction

**地点** Bargino, San Casciano Val di Pesa，佛罗伦萨，意大利

**项目** 酿酒厂、办公楼

**客户** Marchesi Antinori s.r.l.

**项目管理和工程设计** Hydea's.r.l., Paolo Giustiniani

**结构设计** A&I progetti s.r.l., Massimo Tom, Niccolò De Robertis

**系统设计** M&E s.r.l., Stefano Mignani, Paolo Bonacorsi

**葡萄栽培装置** Emex Engieneering S.r.l. Trading-pty-L.t.d.

**规划** 2004年-2008年

**建设** 在建

**成本** 65,000,000欧元

**占地面积** 139,950平方米

**建筑面积** 41,165平方米

**体量** 287,260立方米

**承包商** Inso S.p.a.

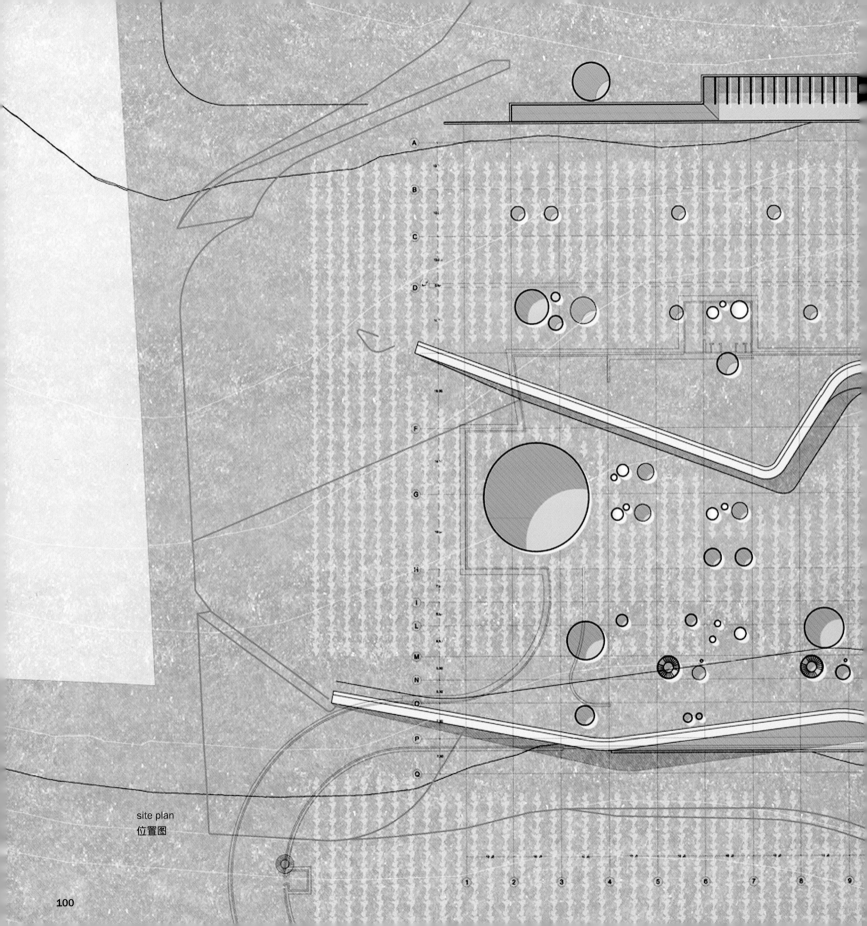

site plan
位置图

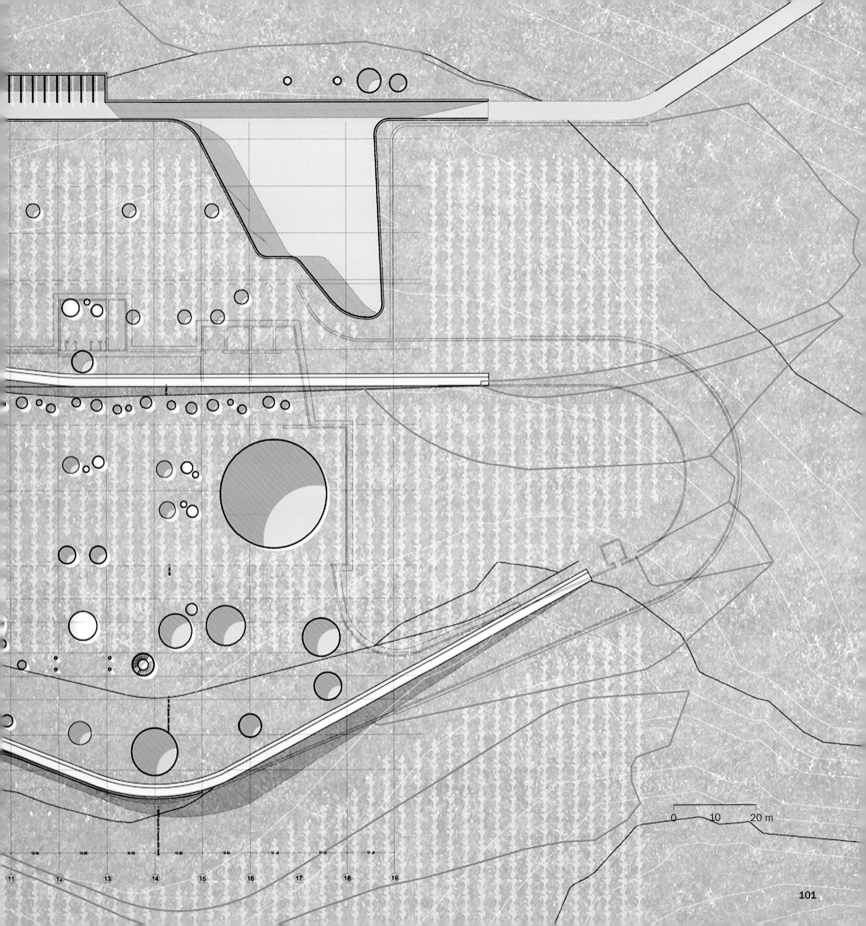

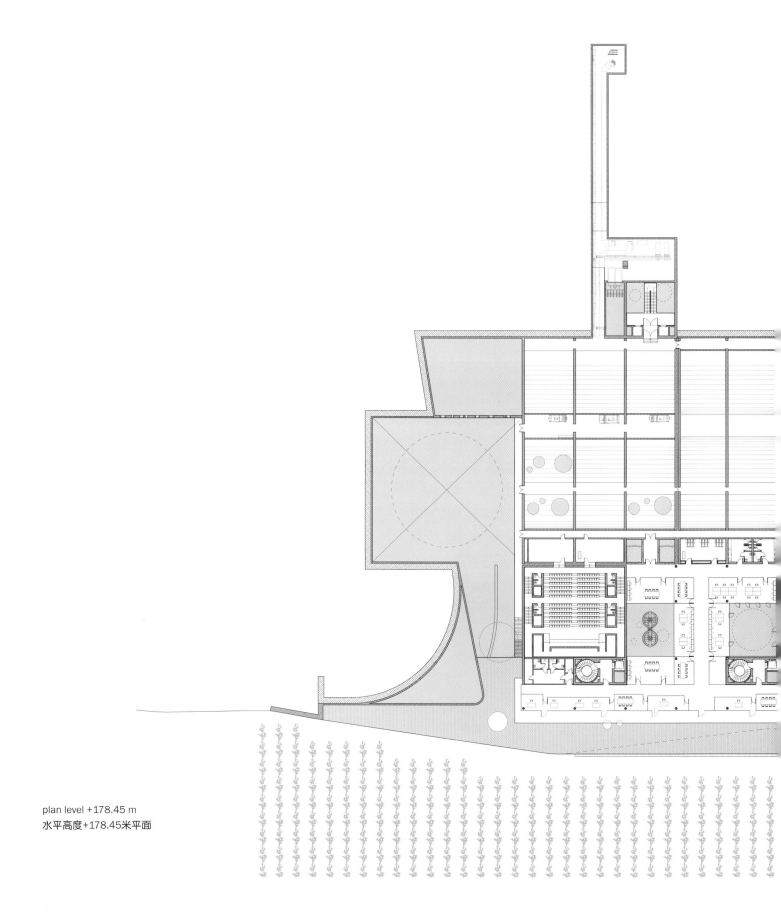

plan level +178.45 m
水平高度+178.45米平面

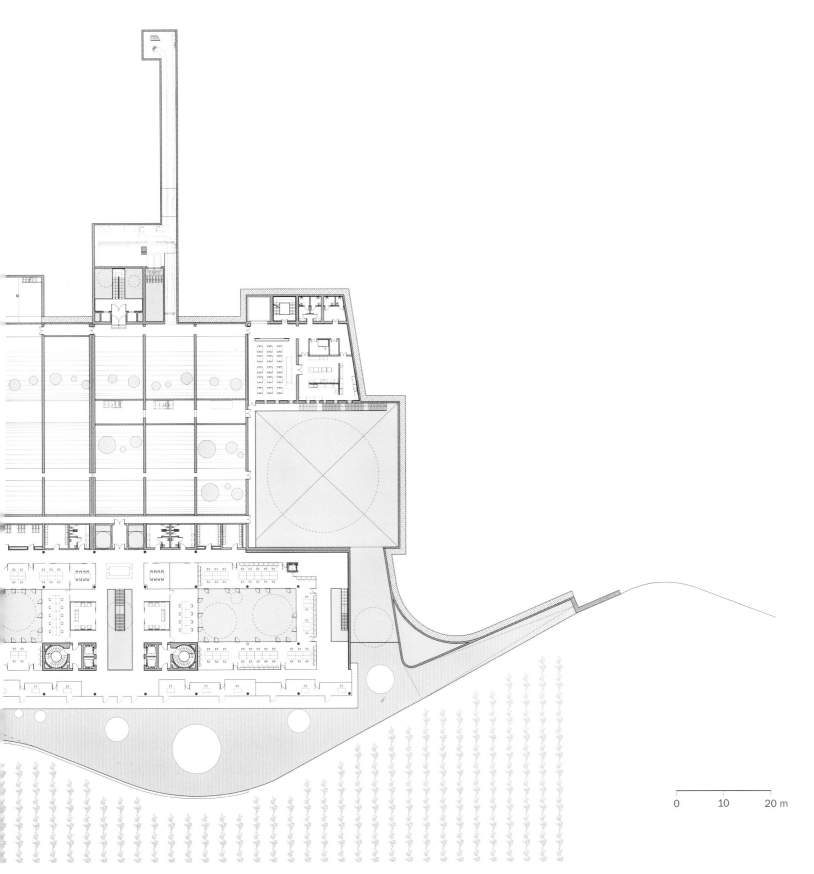

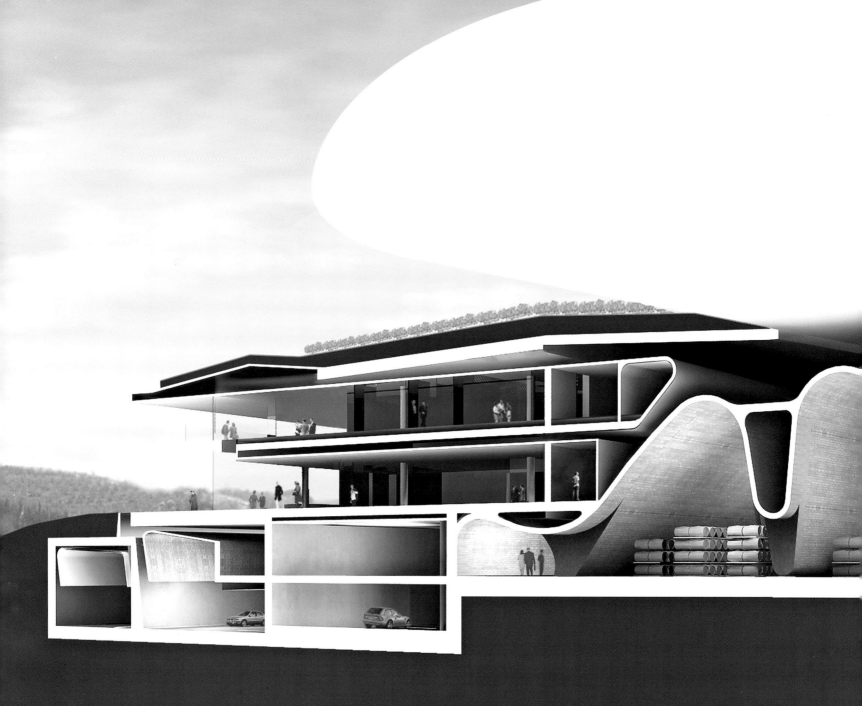

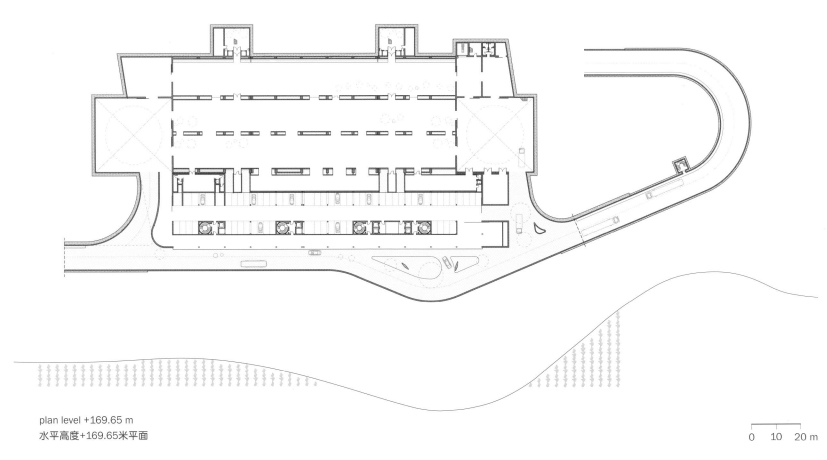

plan level +169.65 m
水平高度+169.65米平面

0　10　20 m

# TIANJIN LAND 7

## office building
## Tianjin, China, 2005

地点　中国天津
项目　管理和商业
客户　HEDO Construction
and Investment Group
结构　Favero&Milan Ingegneria
规划　2005 年竞赛，第2名
成本　15,000,000欧元
建筑面积　18,000平方米
体量　90,000立方米

site plan
位置图

0  20    50 m

longitudinal section
纵向剖面图

ground floor plan
一层平面图

0 5 10 m

# MASTERPLAN OF CASTELLO

## Florence, Italy, 2005-2007

| 地点 | Castello, Florence, Italy |
| --- | --- |
| 项目 | Residential, commercial, directional |
| 客户 | Europrogetti S.r.l. |
| 规划 | 2005年-2007年 |
| 景观项目 | Atelier Girot |
| 占地面积 | 1,200,000平方米 |
| 公园面积 | 800,000平方米 |

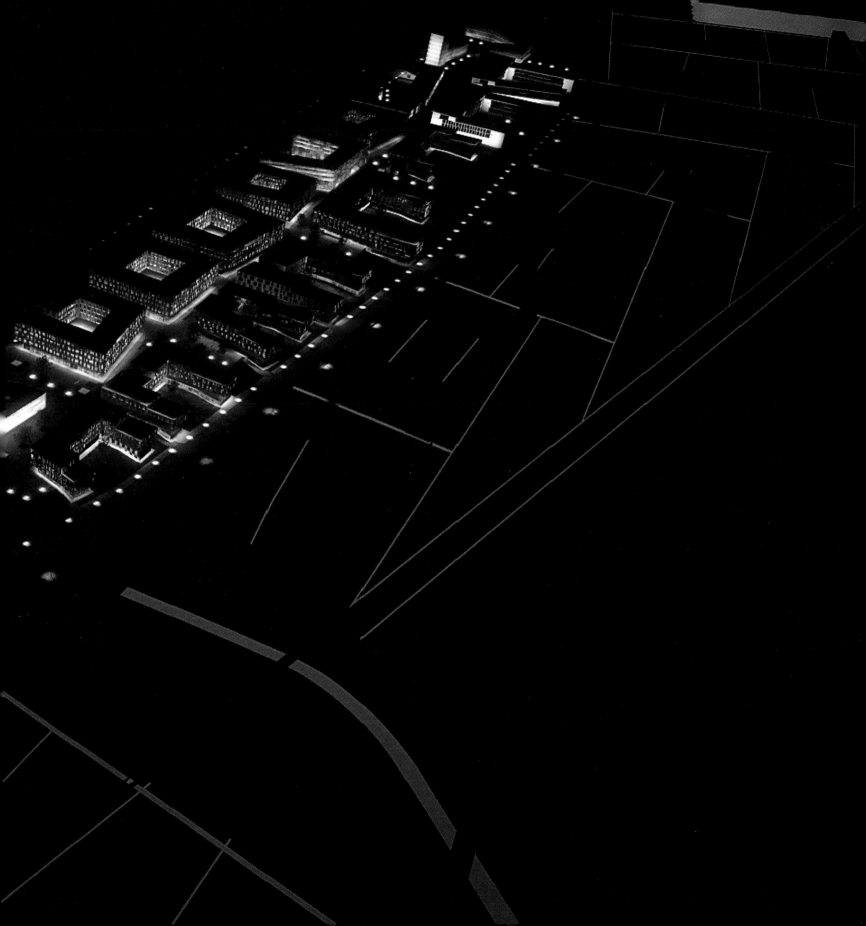

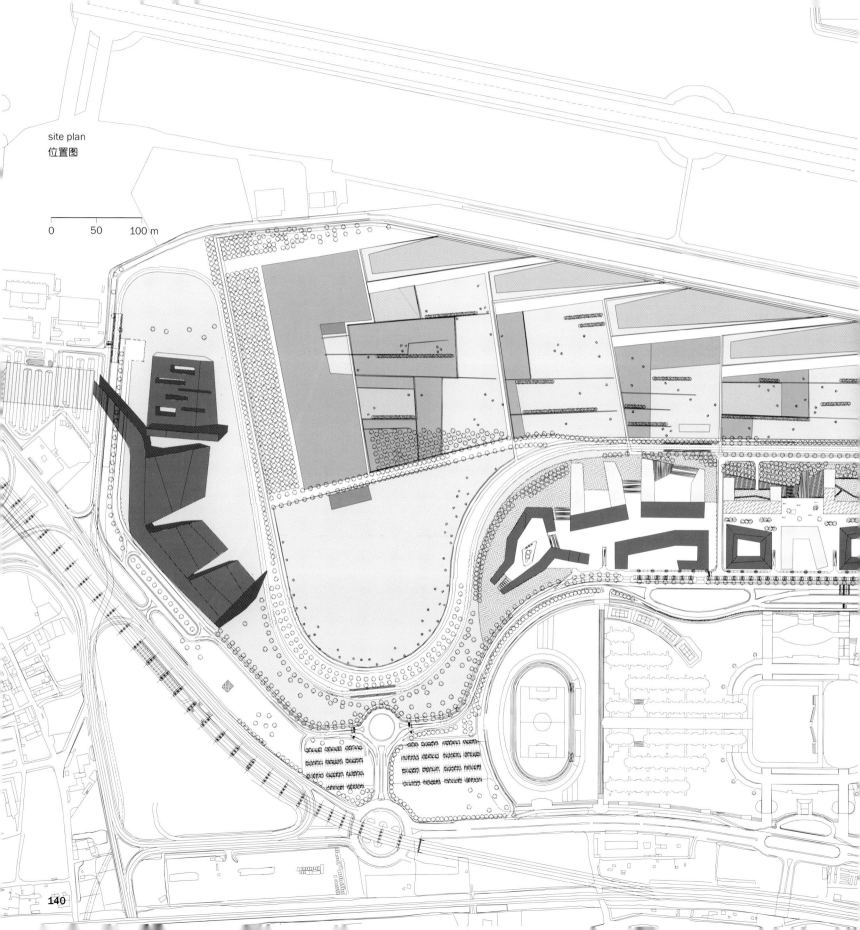

site plan
位置图

0    50    100 m

141

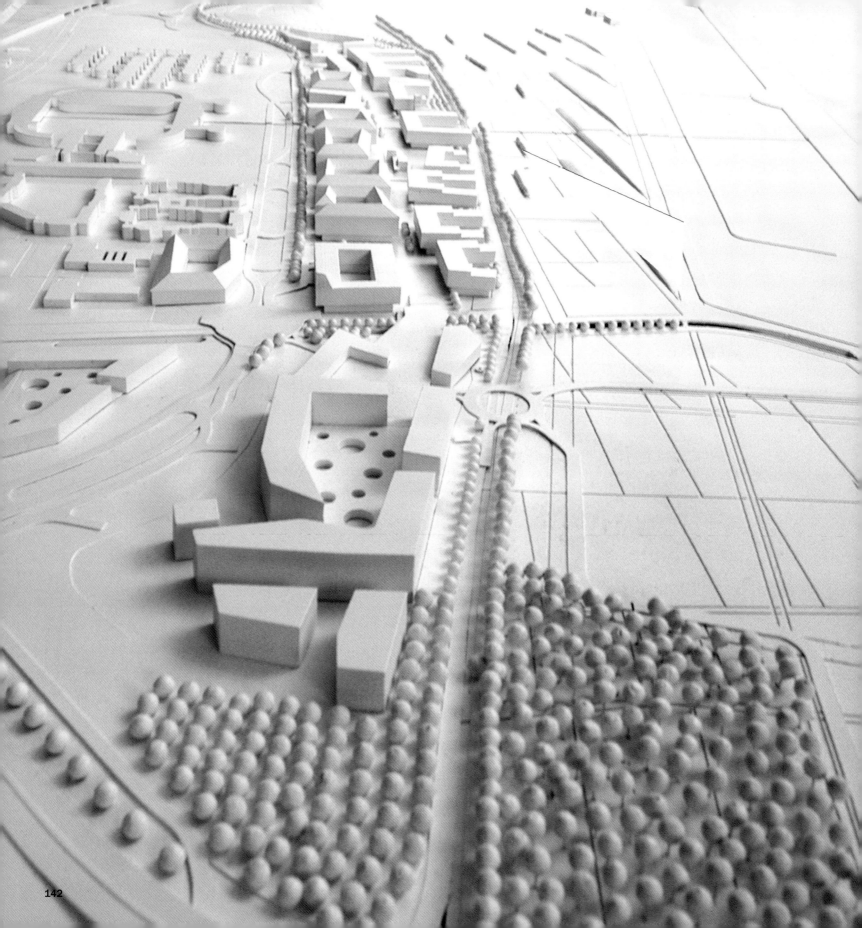

# MASTERPLAN O CASTELLO REGIONAL GOVERNMENT BUILDING

## Florence, Italy, 2005-2007

| | |
|---|---|
| 地点 | Castello, Florence, Italy |
| 项目 | Residential, commercial, directional |
| 客户 | Europrogetti S.r.l. |
| 规划 | 2005年-2007年 |
| 景观项目 | Atelier Girot |
| 占地面积 | 1,200,000平方米 |
| 公园面积 | 800,000平方米 |

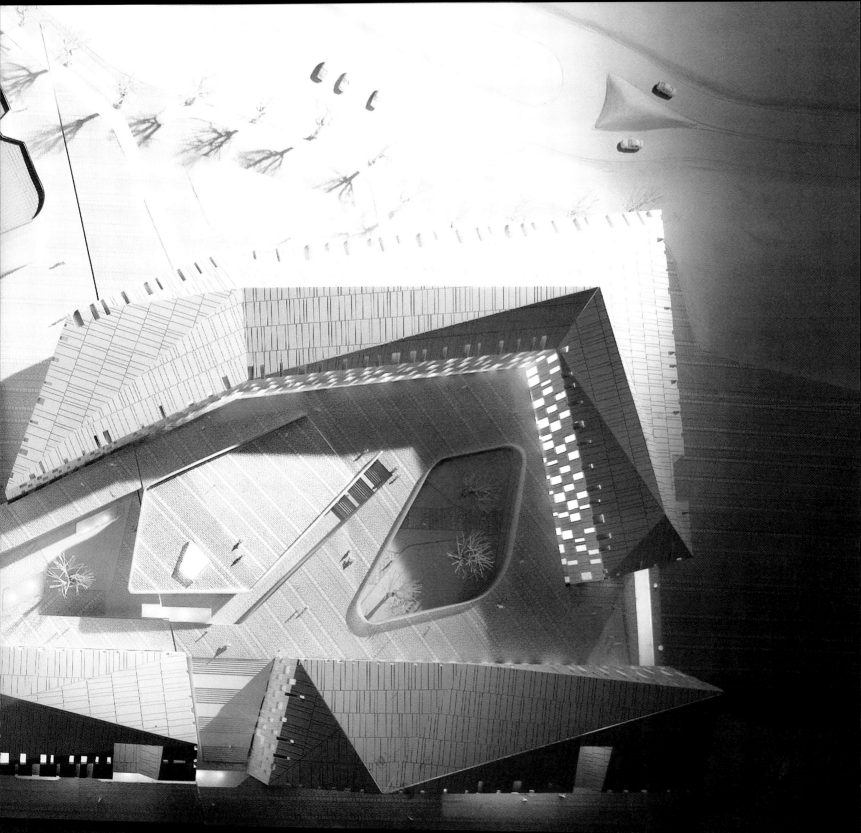

elevations
立面

0    10    20 m

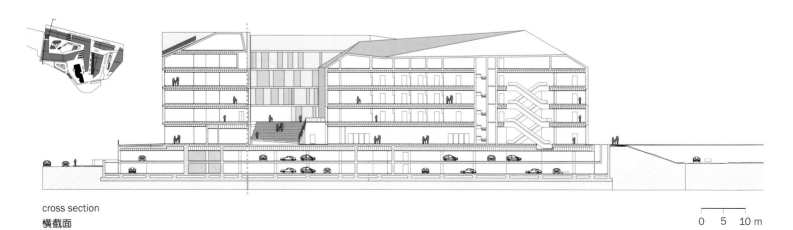

cross section
横截面

0    5    10 m

ground floor plan
一层平面图

0   5   10 m

third floor plan
四层平面图

0  5  10 m

# MASTERPLAN OF CASTELLO DISTRICT GOVERNMENT BUILDING

## Florence, Italy, 2005-2007

**地点**　Castello，佛罗伦萨，意大利

**项目**　办公楼

**客户**　Europrogetti S.r.l.

**系统**　Hilson&Moran

**规划**　2005年-200年

**成本**　50,000,000欧元

**建筑面积**　20,000平方米

**体量**　70,000立方米

154

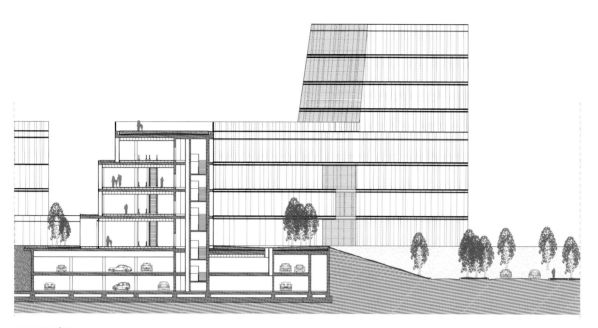

cross section
横截面

ground floor plan
一层平面图

0    10    20 m

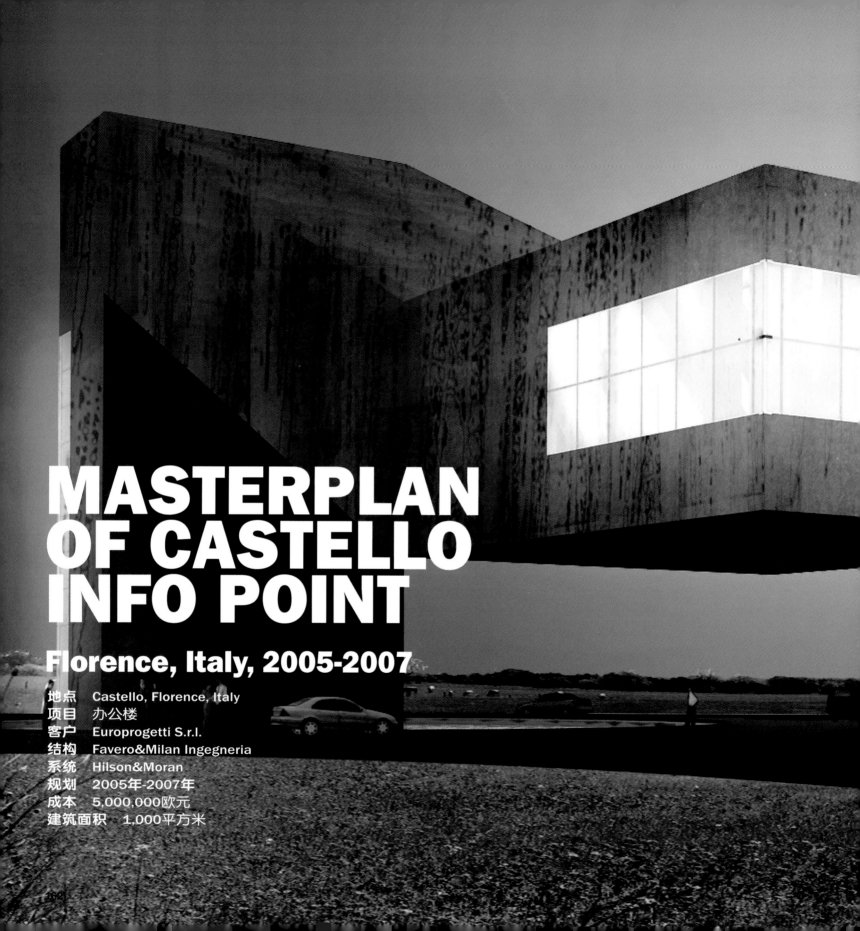

# MASTERPLAN OF CASTELLO INFO POINT

## Florence, Italy, 2005-2007

地点　Castello, Florence, Italy
项目　办公楼
客户　Europrogetti S.r.l.
结构　Favero&Milan Ingegneria
系统　Hilson&Moran
规划　2005年-2007年
成本　5,000,000欧元
建筑面积　1,000平方米

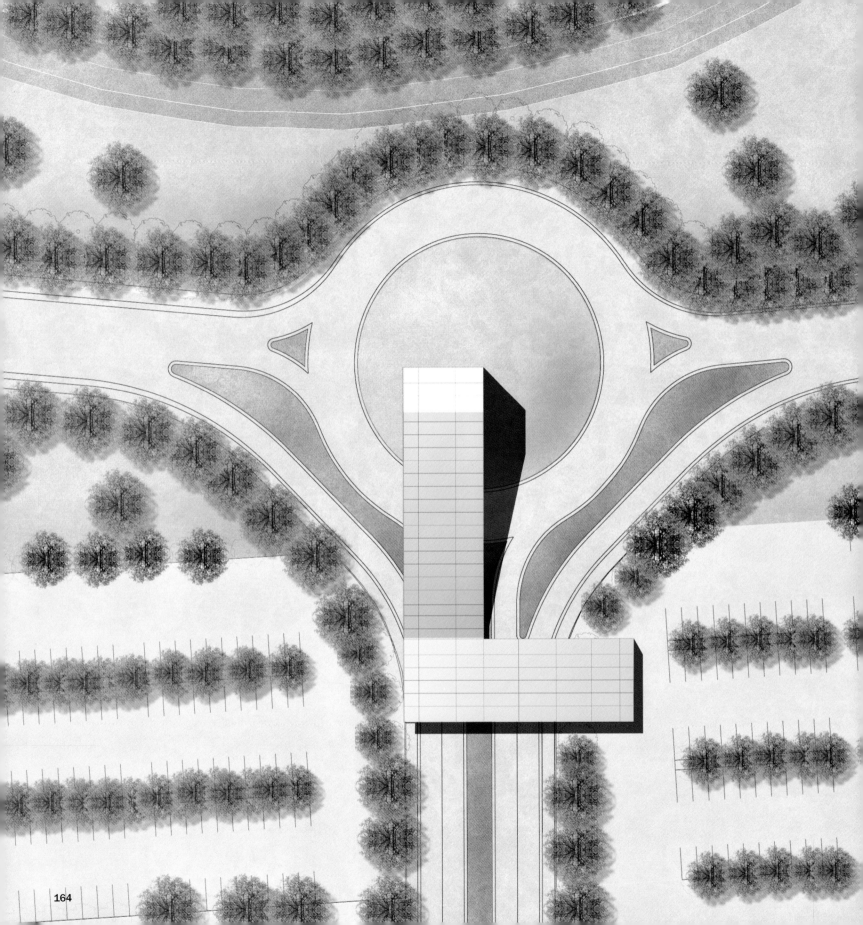

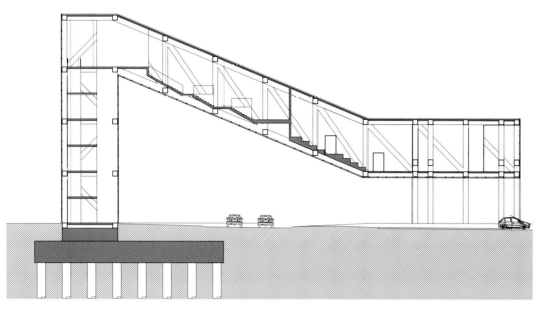
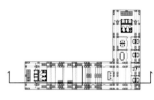

longitudinal section
纵剖面

0    5    10 m

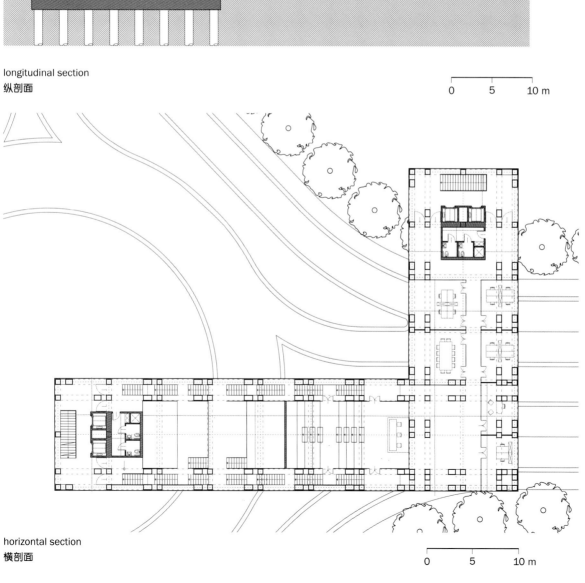

horizontal section
横剖面

0    5    10 m

# HIGHWAY MUSEUM

## Salerno - Reggio Calabria, Italy, 2006

地点　萨勒诺与勒佐卡拉布里亚间
高速公路，意大利
项目　景观恢复和高速公路博物馆
客户　ANAS S.p.A.
结构　Giuliano Sauli
系统　Sistemi Industriali S.r.l.
规划　2006年设计竞赛，第一名
成本　20,000,000欧元
绿地　15,000平方米
停车面积　55,820平方米

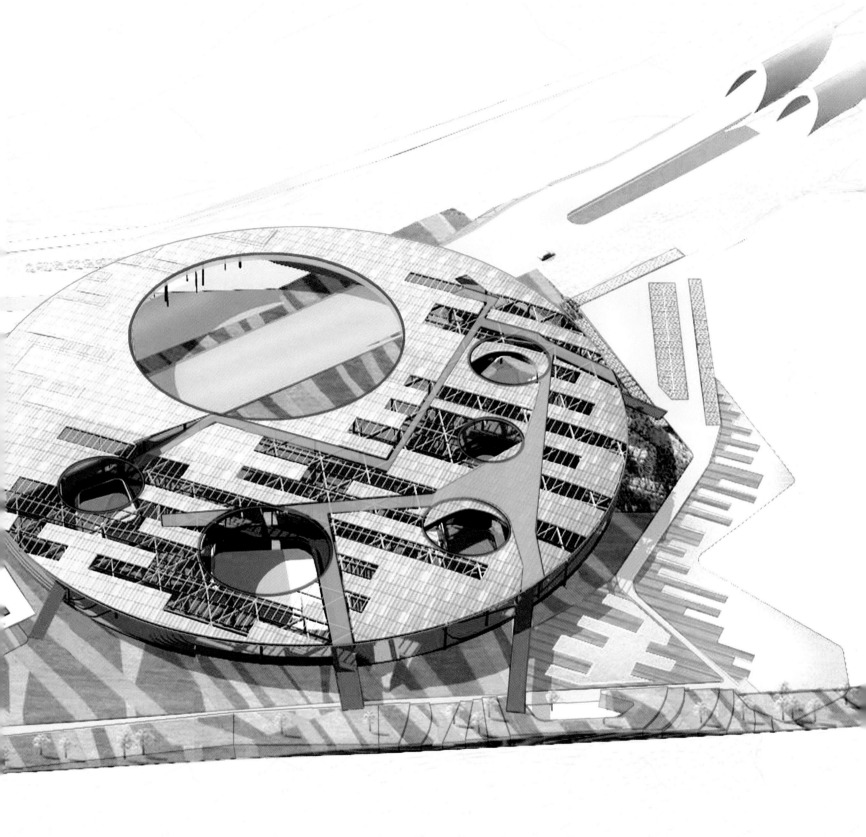

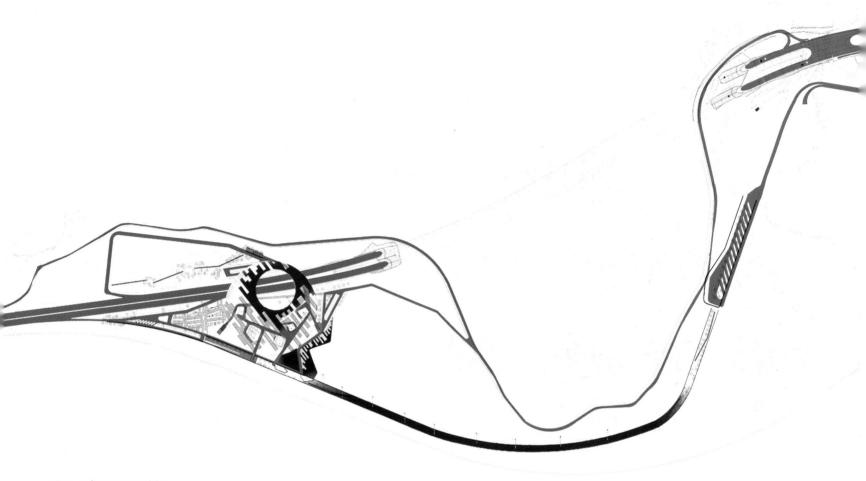

scheme of energy supplying
能源供应方案

section on the archaeological museum
考古博物馆立面

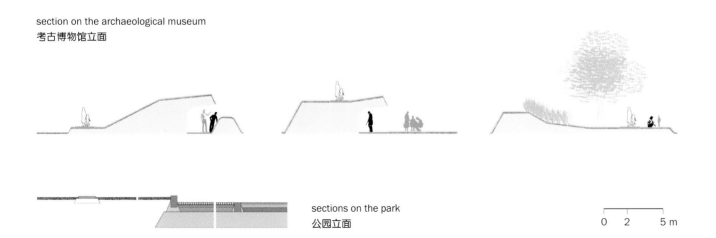

sections on the park
公园立面

0  2    5 m

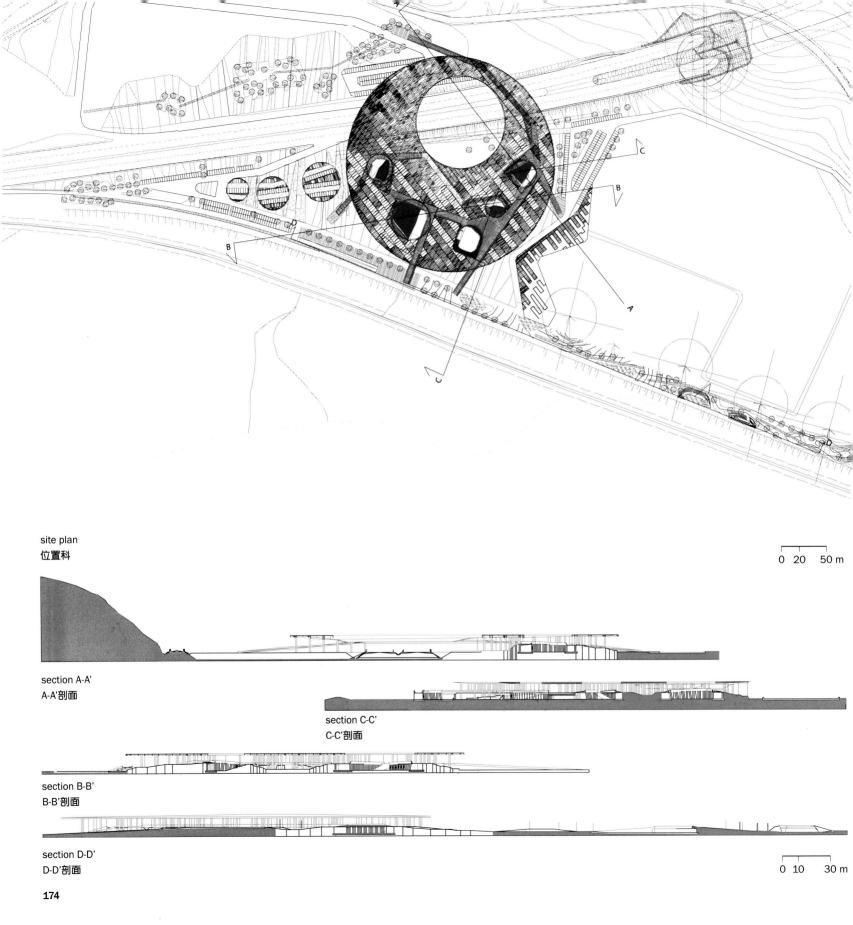

site plan
位置科

0  20  50 m

section A-A'
A-A'剖面

section C-C'
C-C'剖面

section B-B'
B-B'剖面

section D-D'
D-D'剖面

0  10  30 m

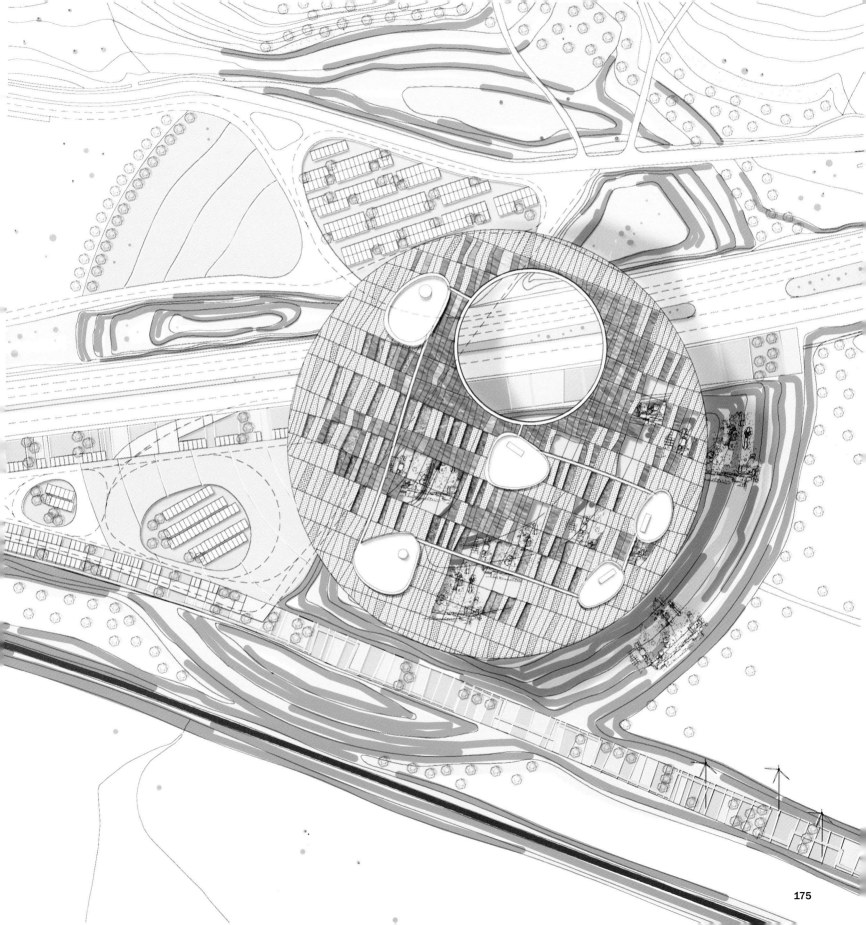

# TICOSA EX-INDUSTRIAL AREA
## Como, Italy, 2006-2007

地点　前Ticosa纺织区区域，科摩，意大利
项目　住宅和商业
客户　Multi Veste Italy S.r.l.
结构　MS Progetti
系统　Studio Pedrini
规划　2006年 -2007年
成本　60,000,000欧元
绿地　15,000平方米
停车面积　55,820平方米
建筑面积　39,649平方米

the relationship with existing historical structures
与现有历史建筑的关系

main connecting commercial squares
主要连接的商业广场

main connecting commercial squares
主要连接的商业广场

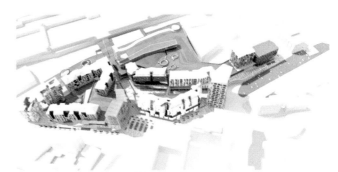

the green areas in the new Ticosa park
提卡索公园的绿色区域

179

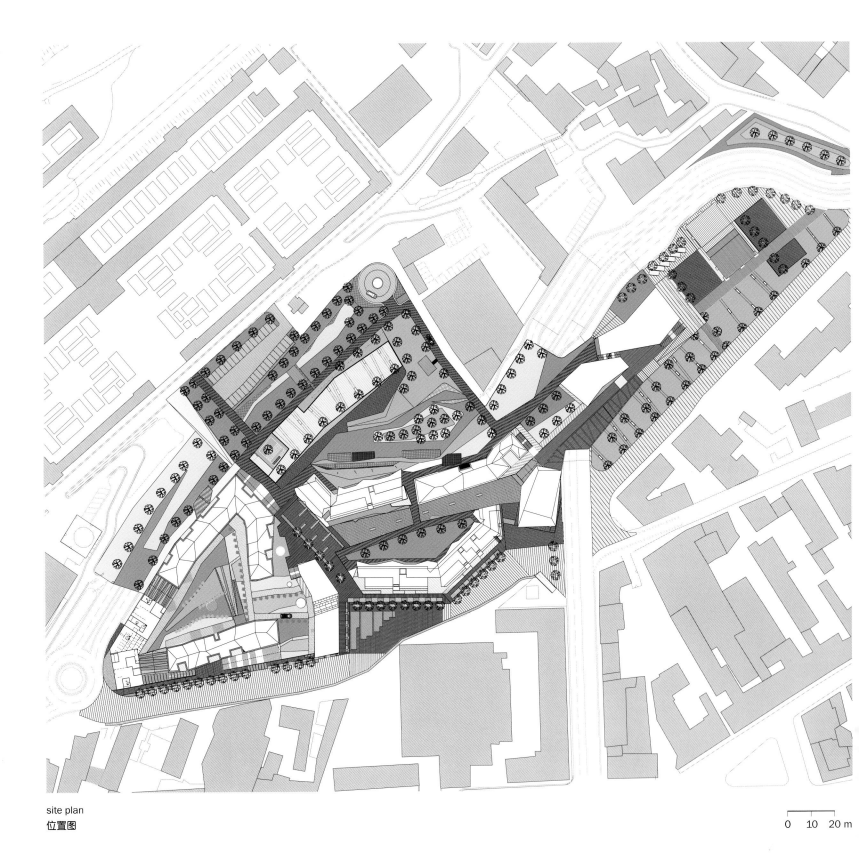

site plan
位置图

0 10 20 m

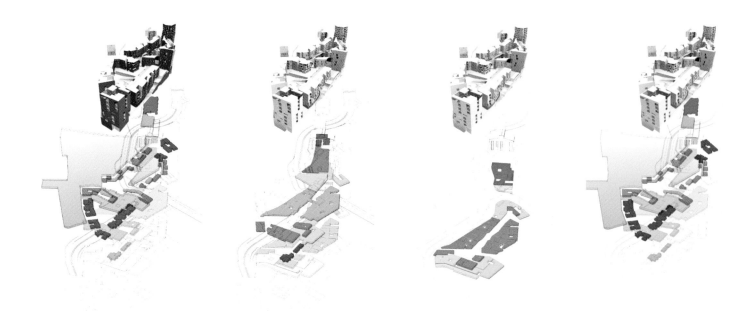

functional schemes
功能配置方案

- commercial / 广告
- neighbourhood commercial / 附近的广告
- fitness / 健身
- shops / 商店
- residences / 居住
- receptive areas / 接待区
- directional areas / 指示区
- services / 服务

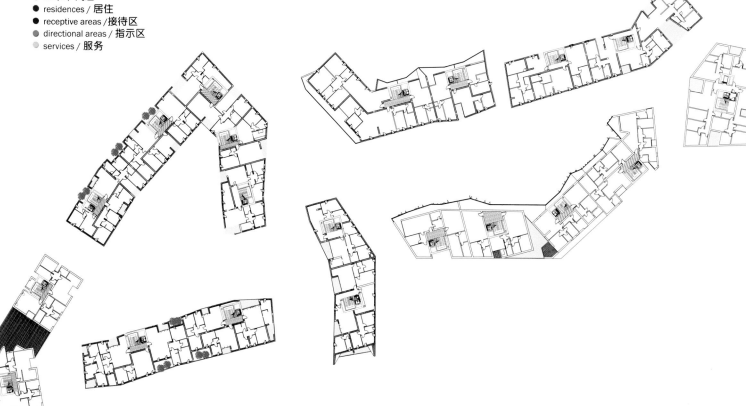

type plan
位置图

0  5  10 m

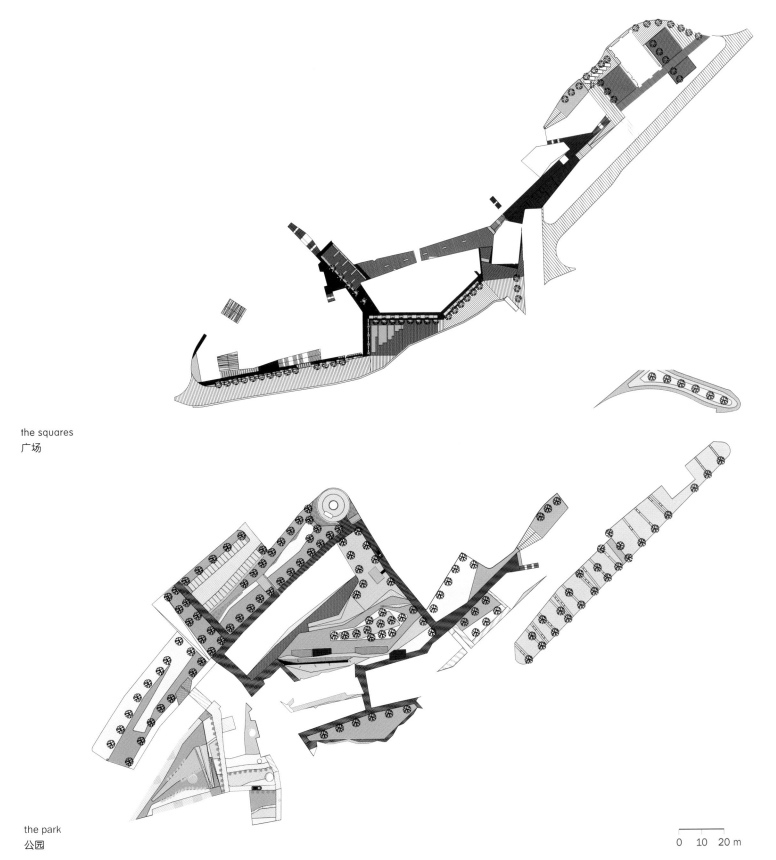

the squares
广场

the park
公园

0   10   20 m

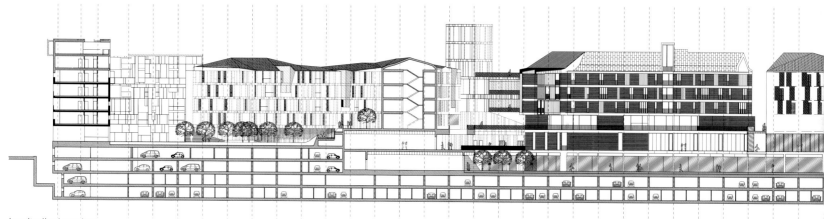

longitudinal section
纵剖面

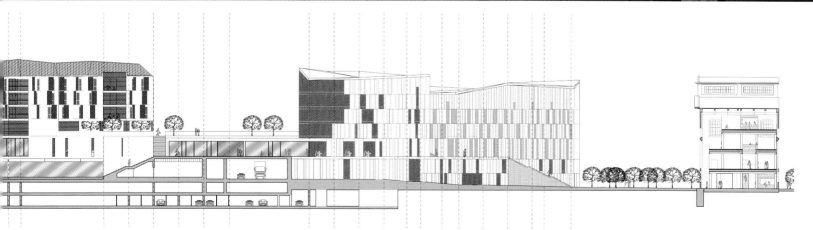

0    5    10 m

longitudinal section
纵剖面

# BORGO ARNOLFO

## residential and commercial complex
## San Giovanni Valdarno, Florence, Italy
## 2006 - under construction

地点　San Giovanni Valdarno，佛　　规划　2004年-2008年
罗伦萨，意大利　　　　　　　　　　建设　在建
项目　住宅楼、商用楼　　　　　　成本　20,000,000欧元
客户　Etruria Investimenti S.p.A.　面积　7,686平方米
结构　Studio Bacci & Bandini　　　体量　25,365立方米
ingegneri associati　　　　　　　　承包商　Etruria Investimenti
系统　Giuliano Galzigni　　　　　　S.c.a.r.l.
隔音　Annalisa Baracchi

ground floor plan
一层平面图

190

first floor plan
二层平面图

0   5   10 m

192

third floor plan
四层平面图

# ALBATROS CAMPING

## San Vincenzo, Livorno, Italy, 2006-2009

地点　San Vincenzo，里窝那，意大利
项目　酒店设施
客户　Park Albatros s.a.s.
结构　Roberto Nocentini
系统　Leonardo Bracciali
电气系统　S.T.E. società toscana elettrica S.r.l.
泳池系统　Acqua Sport Service S.r.l.
规划　2006年
建设　2007年-2009年
成本　20,000,000欧元
占地面积　40公顷
公共卫生区块面积　1,100平方米
市场面积　1,650平方米
承包商　T.I.S.

site plan
位置图

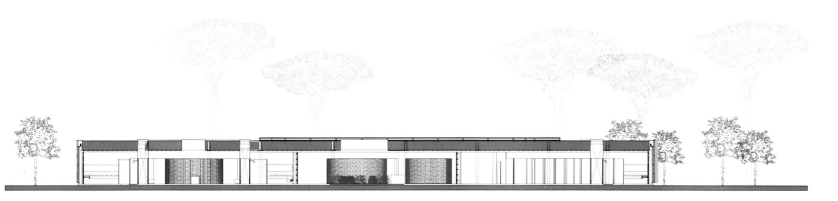

longitudinal section service area
服务区纵向剖面

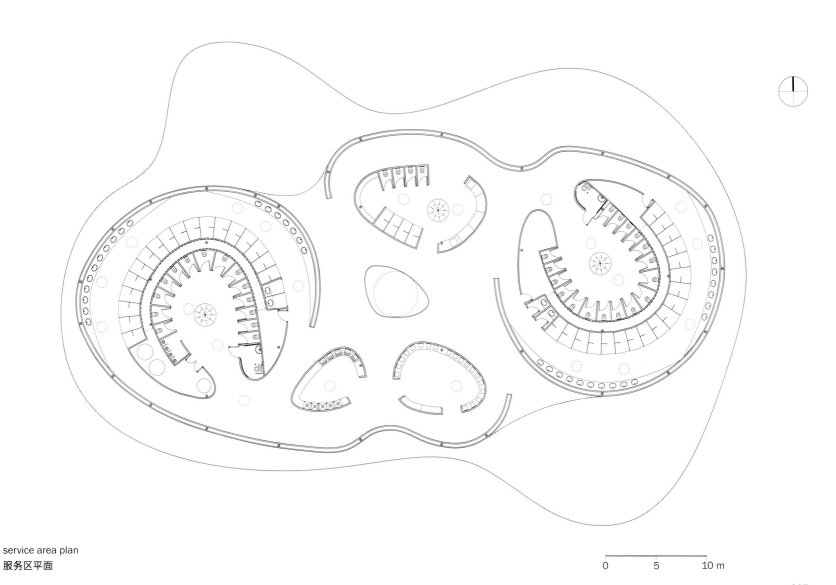

service area plan
服务区平面

0    5    10 m

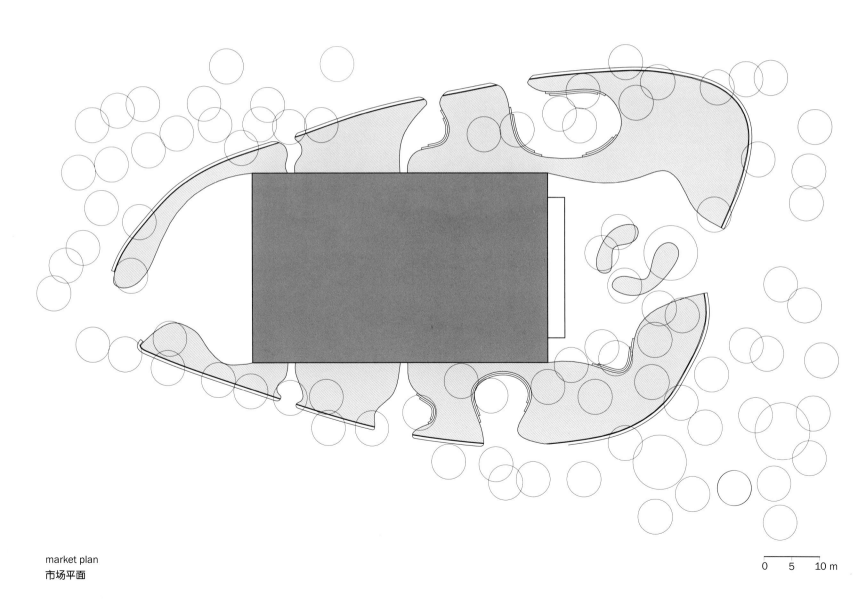

market plan
市场平面

0    5    10 m

# PARKOUR-BEIJING

## Beijing, China, 2007

地点　北京，中国
项目　住宅和商业
客户　Beijing Xisi-Bei Ltd.
规划　2007年
建筑面积　18,000平方米
体量　200,000立方米

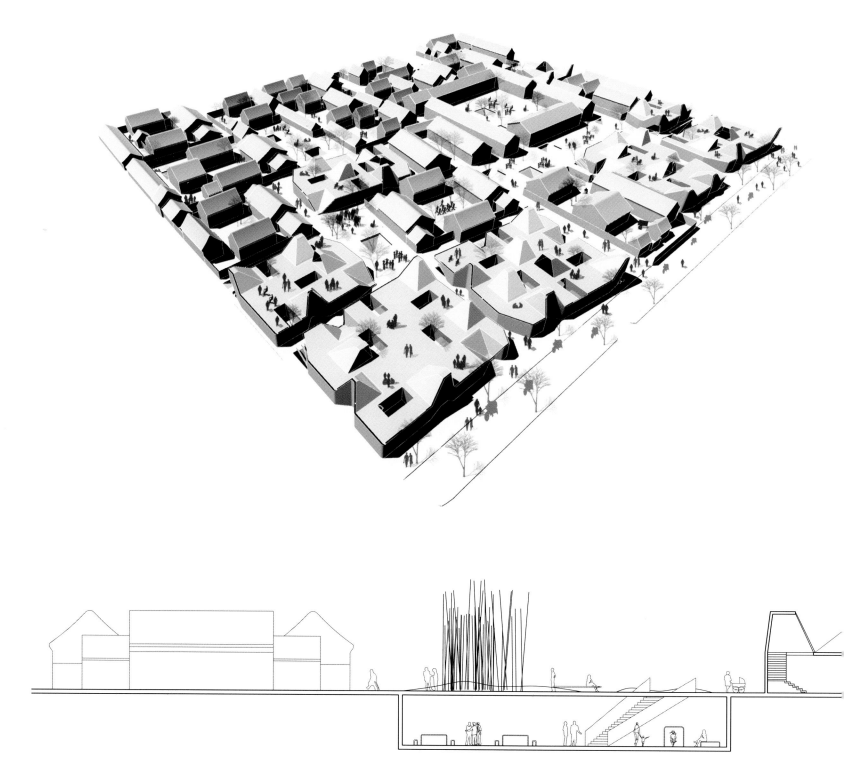

cross section
横剖面

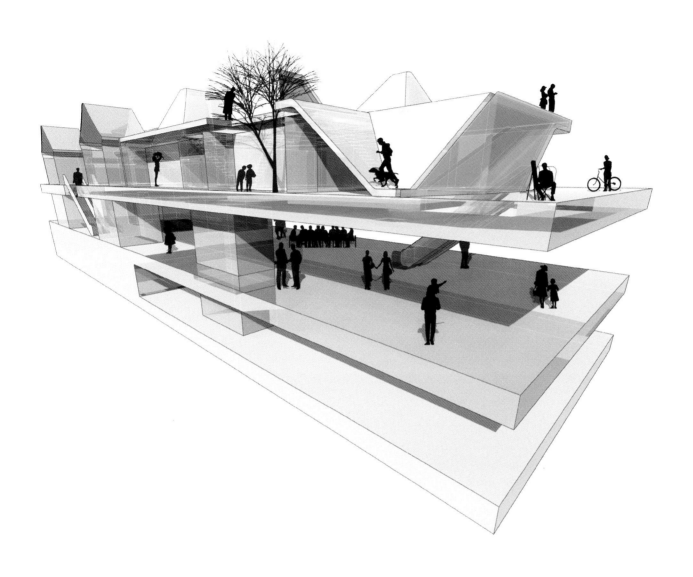

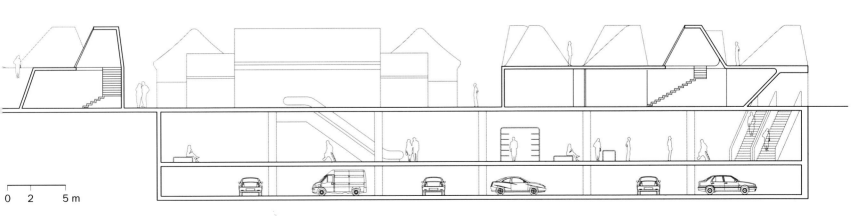

0  2    5 m

# SHANGRI-LA WINERY

## Penglai, China, 2007 - under construction

**地点** 蓬莱，中国
**项目** 工业
**客户** Shangri-la Winery Ltd.
**规划** 2007 年
**占地面积** 80,000平方米
**建筑面积** 18,000平方米
**体量** 72,000立方米

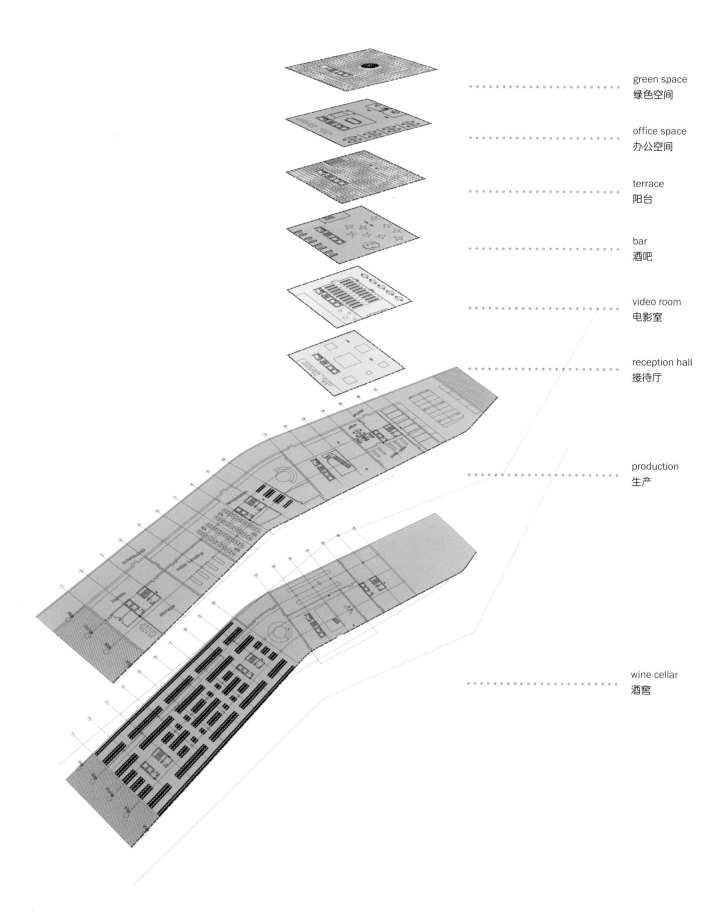

green space
绿色空间

office space
办公空间

terrace
阳台

bar
酒吧

video room
电影室

reception hall
接待厅

production
生产

wine cellar
酒窖

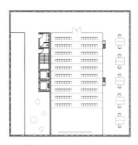

plan video room 18.5 m
电影室平面18.5米

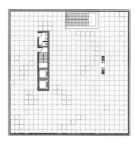

plan terrace 23.5 m
阳台平面23.5米

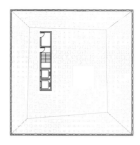

plan green space 37.5 m
绿色空间平面37.5米

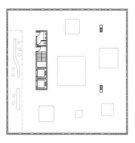

plan reception hall 3.5 m
接待厅平面3.5米

plan bar 18.5 m
酒吧平面18.5米

plan office 23.5 m
办公室平面23.5米

0   5   10 m

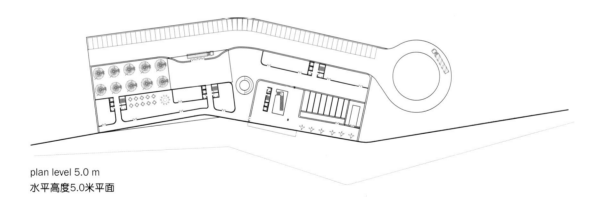

plan level 5.0 m
水平高度5.0米平面

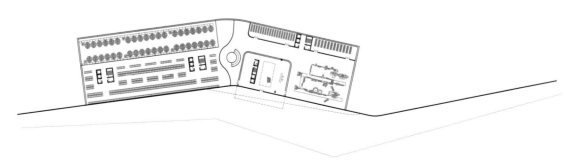

plan level 0.0 m
水平高度0.0米平面

0   10   20 m

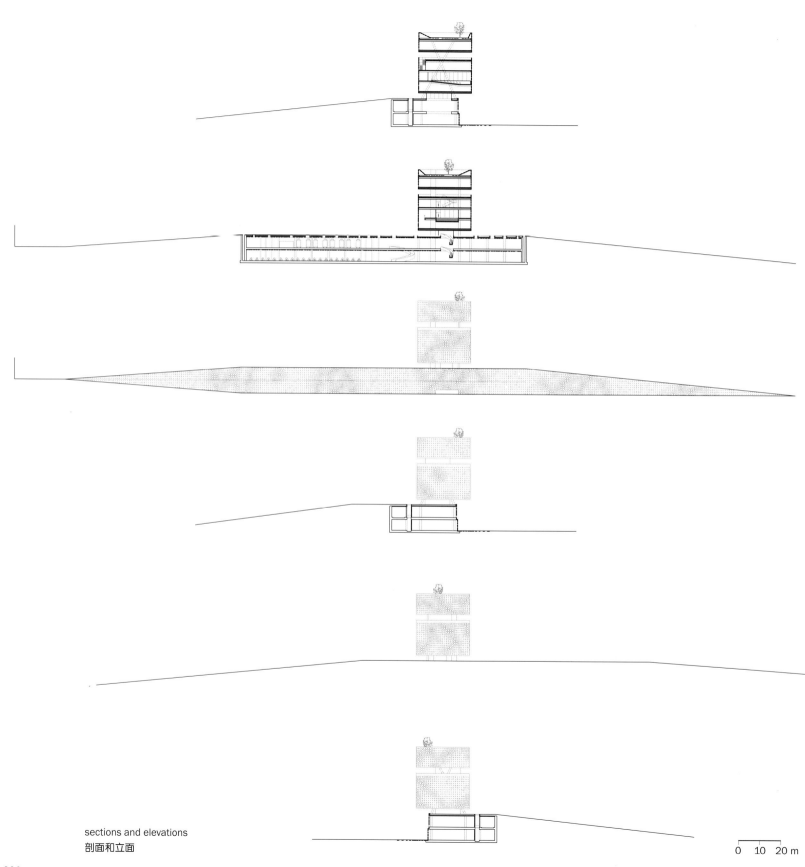

sections and elevations
剖面和立面

0   10   20 m

# EX ITALCEMENTI

**residential and commercial complex**
**Incisa Valdarno, Florence, Italy**
**2008 - in progress**

地点　Incisa Valdarno，佛罗伦萨，意大利
项目　住宅和商业
客户　Siena Est S.r.l
结构　Ariva Engineering
系统　Cianis Progetti
规划　2008年
建设　2010年
成本　15,000,000欧元
总平面面积　18,8
建筑面积　6,850平方
体量　20,636

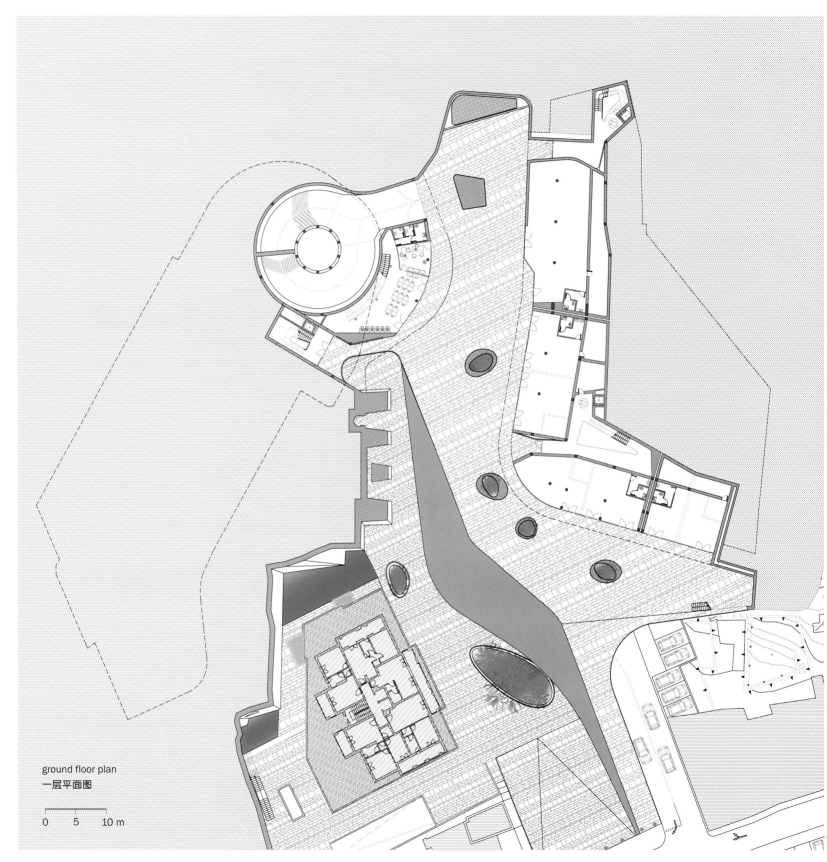

ground floor plan
一层平面图

0   5   10 m

0    5    10 m

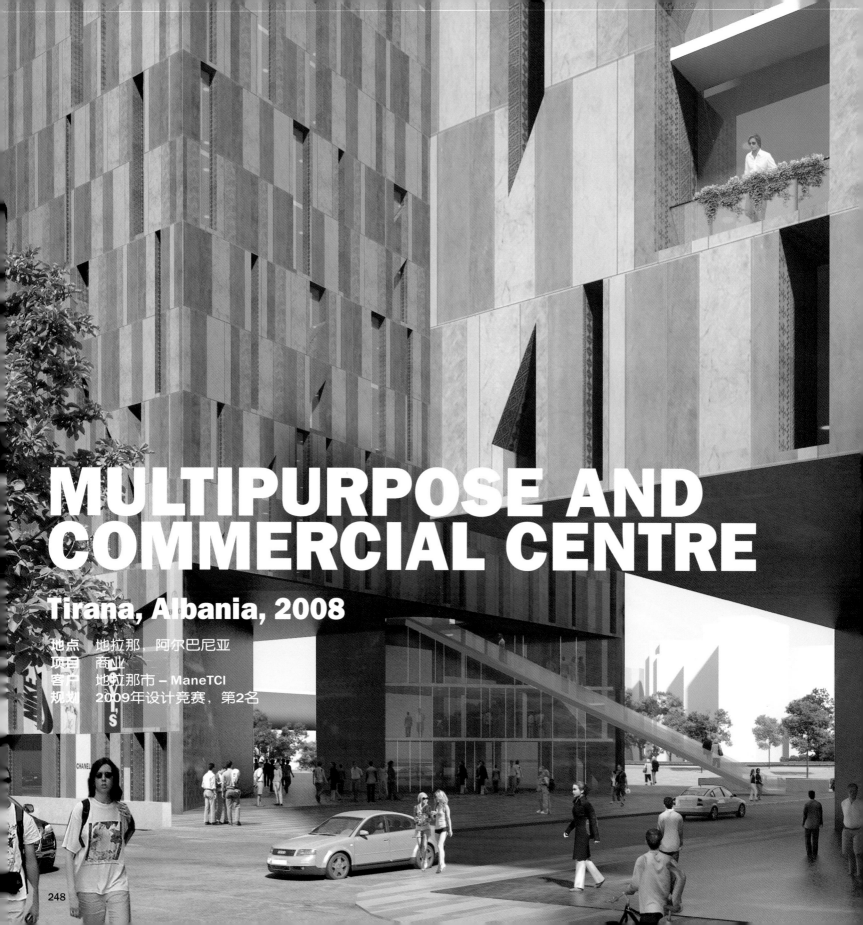

# MULTIPURPOSE AND COMMERCIAL CENTRE

## Tirana, Albania, 2008

地点　地拉那，阿尔巴尼亚
项目　商业
客户　地拉那市 – ManeTCI
规划　2009年设计竞赛，第2名

site plan
位置图

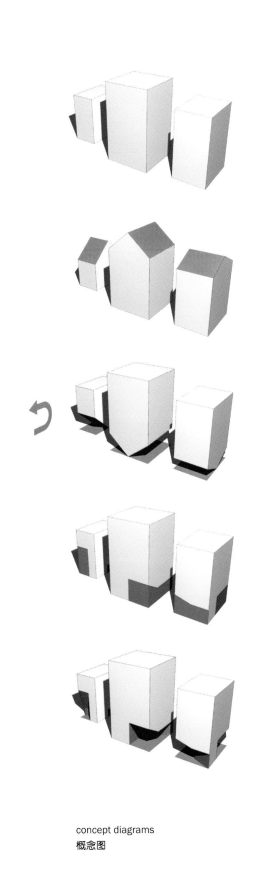

concept diagrams
概念图

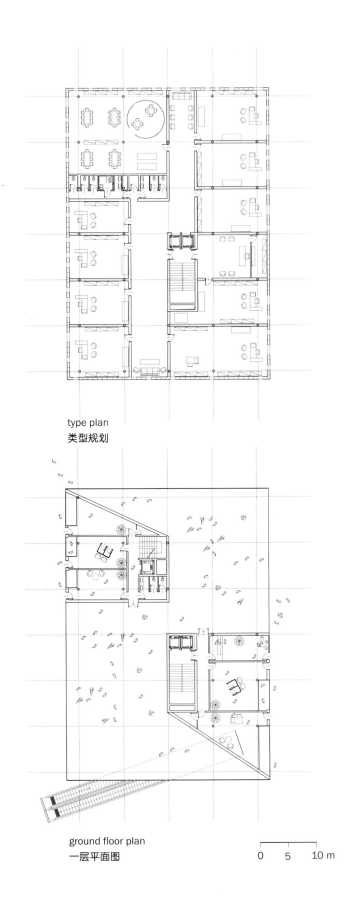

type plan
类型规划

ground floor plan
一层平面图

0　5　10 m

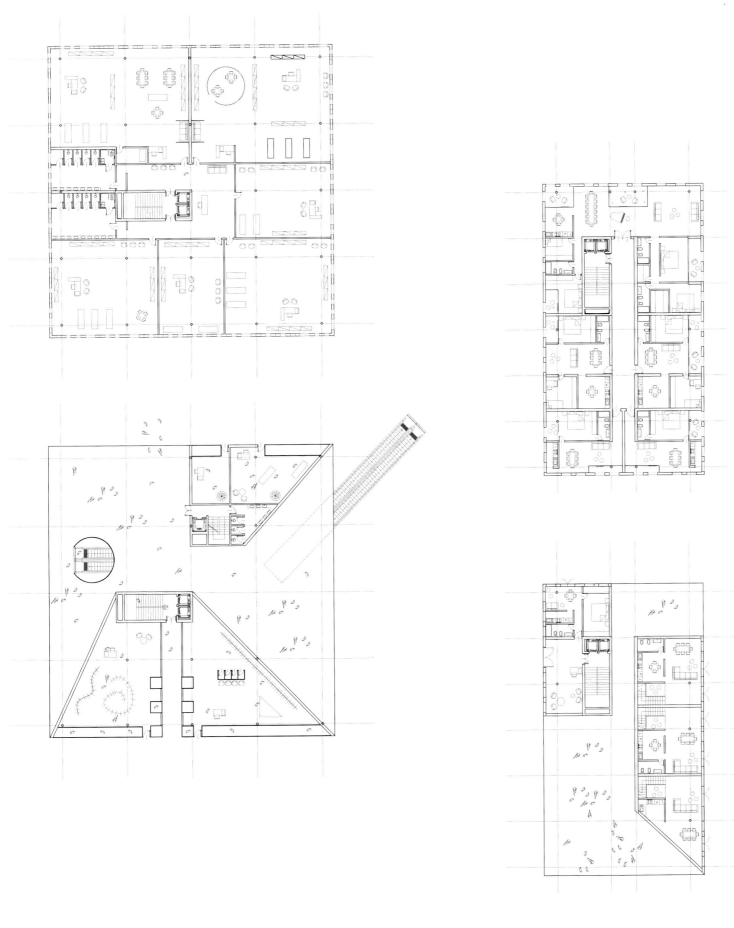

longitudinal section

纵向剖面

0  5  10 m

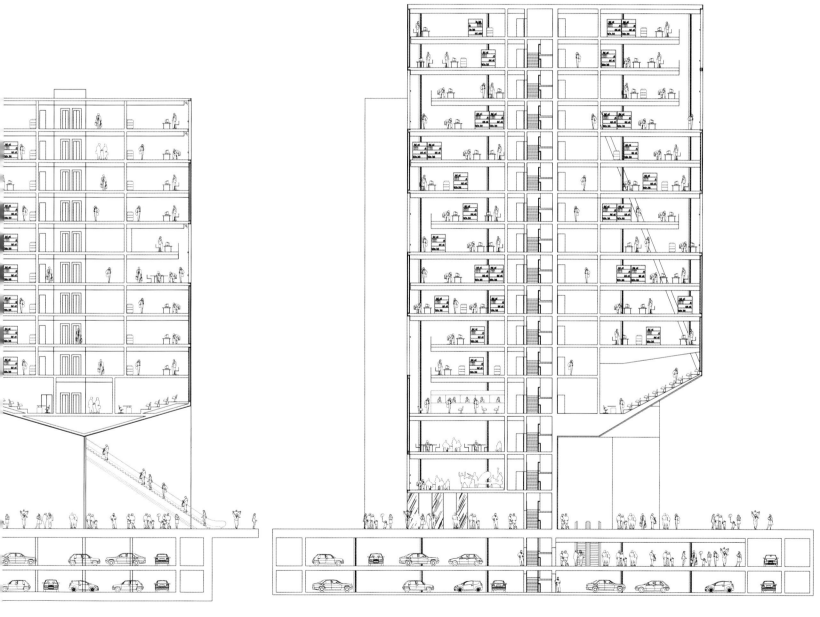

cross section
横截面

# RESIDENZA DEL FORTE CARLO FELICE

hotel and restaurant
La Maddalena, Olbia Tempio, Sassari, Italy
2008-2009

地点　帕多瓦，意大利
项目　总体规划
客户　I.F.I.P.
规划　2008年-2009年
占地面积　61,405平方米
建筑面积　48,506平方米
体量　156,400立方米
承包商　I.F.I.P.

site plan
位置图

0  10  20 m

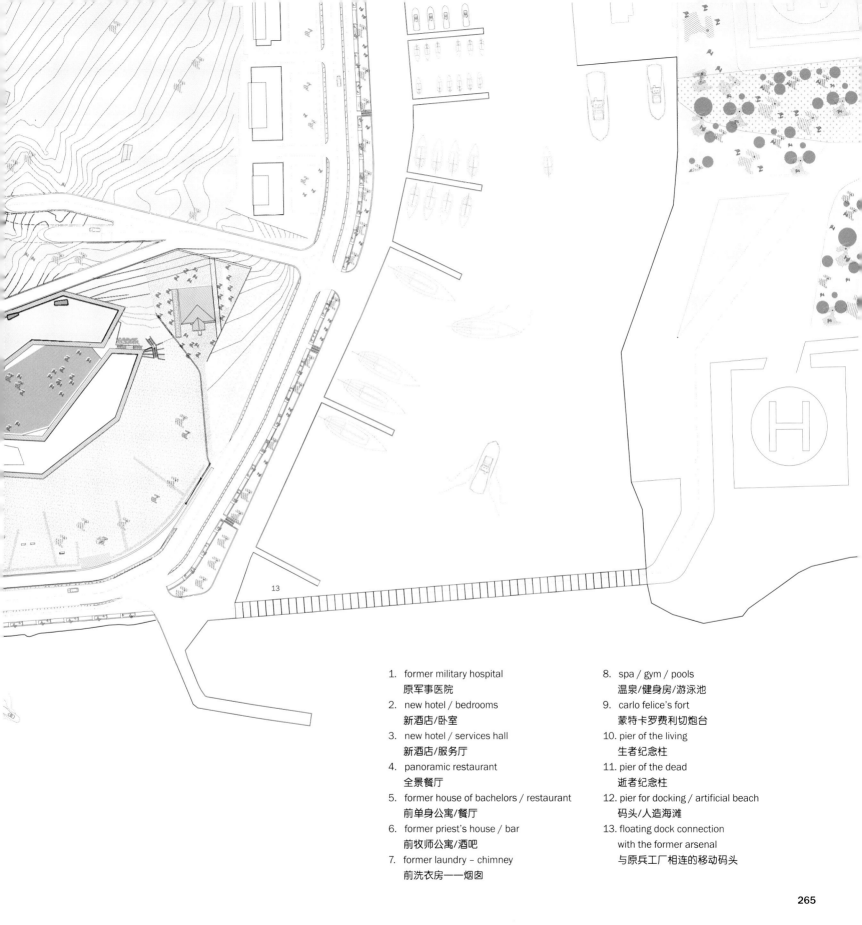

1. former military hospital
   原军事医院
2. new hotel / bedrooms
   新酒店/卧室
3. new hotel / services hall
   新酒店/服务厅
4. panoramic restaurant
   全景餐厅
5. former house of bachelors / restaurant
   前单身公寓/餐厅
6. former priest's house / bar
   前牧师公寓/酒吧
7. former laundry – chimney
   前洗衣房——烟囱

8. spa / gym / pools
   温泉/健身房/游泳池
9. carlo felice's fort
   蒙特卡罗费利切炮台
10. pier of the living
    生者纪念柱
11. pier of the dead
    逝者纪念柱
12. pier for docking / artificial beach
    码头/人造海滩
13. floating dock connection
    with the former arsenal
    与原兵工厂相连的移动码头

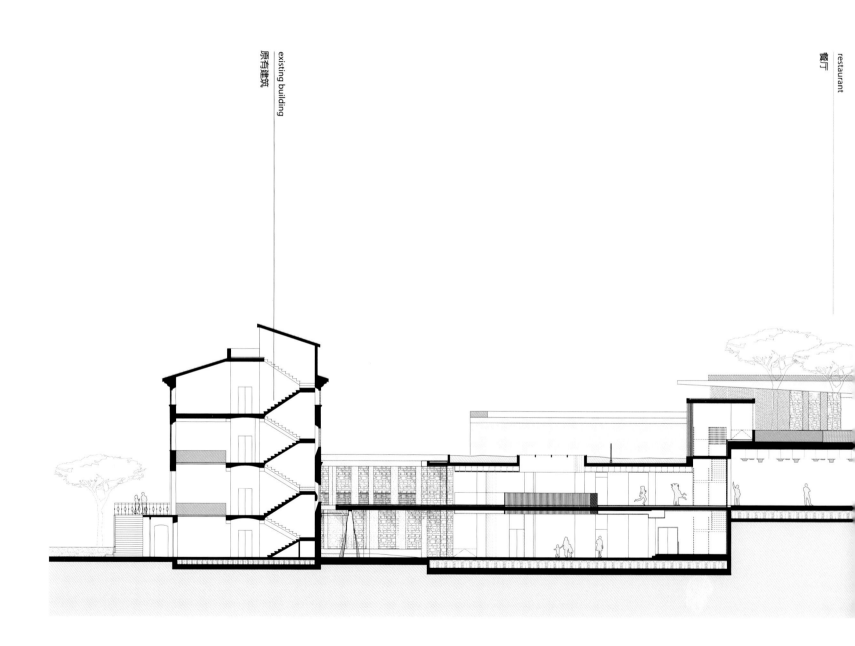

restaurant
餐厅

existing building
原有建筑

cross section
横剖面

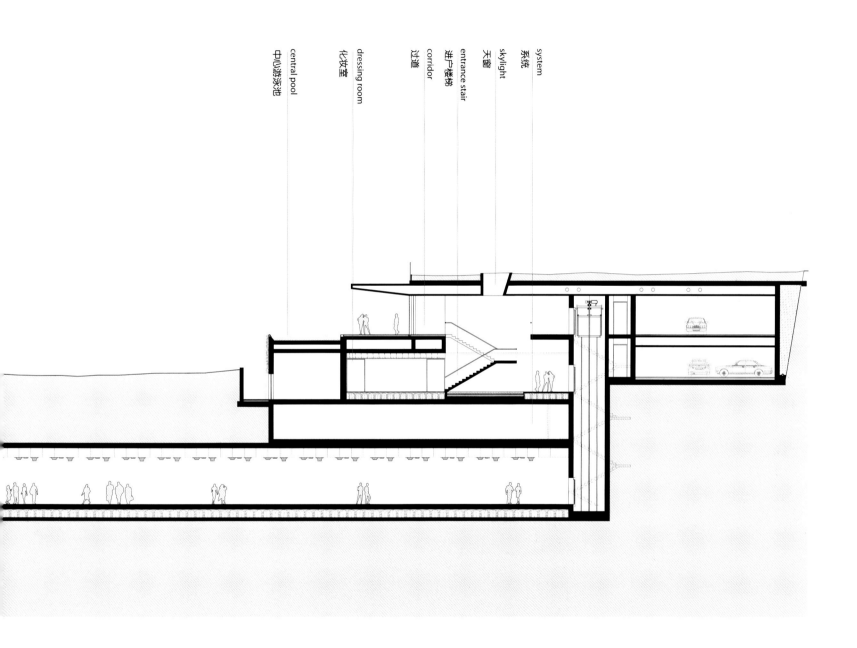

中心游泳池 central pool

化妆室 dressing room

过道 corridor

进户楼梯 entrance stair

天窗 skylight

系统 system

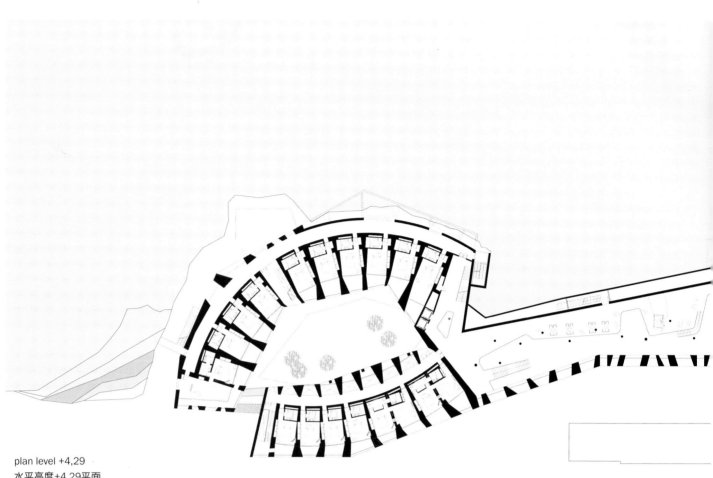

plan level +4,29
水平高度+4.29平面

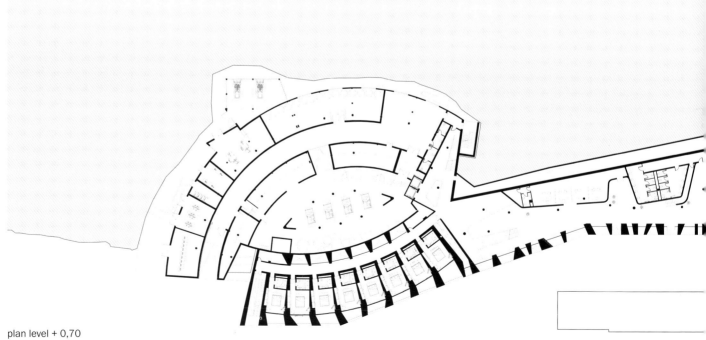

plan level + 0,70
水平高度+0.70平面

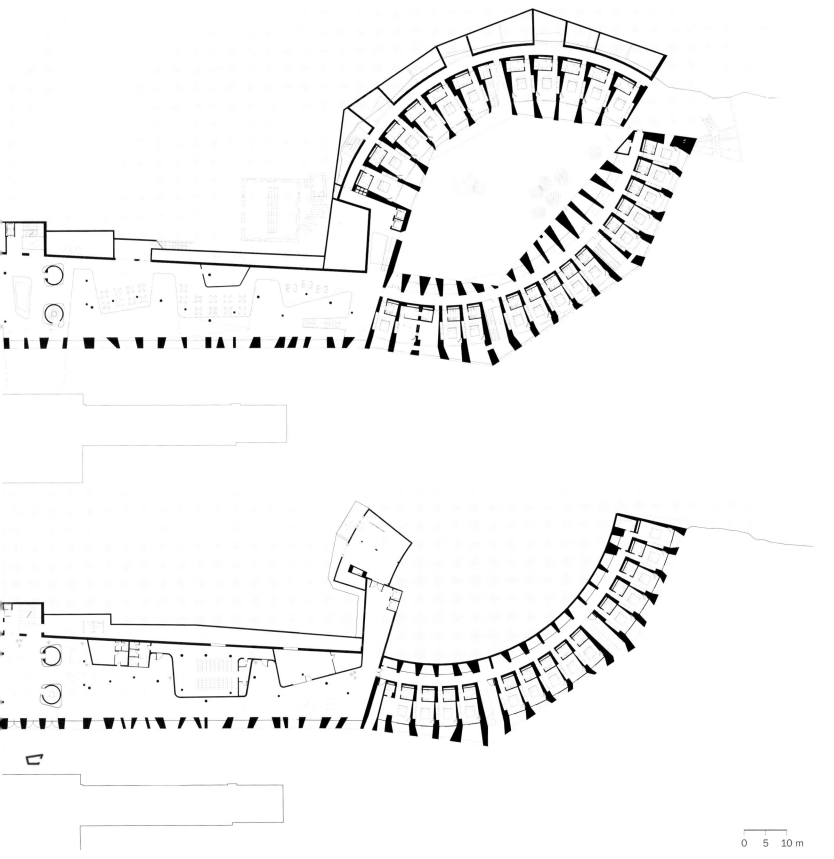

0   5   10 m

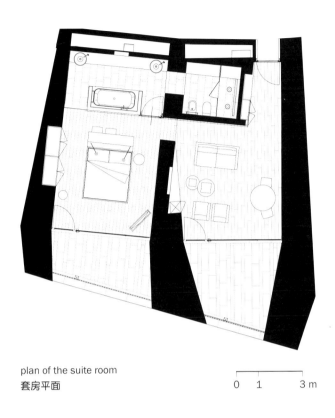

plan of the suite room
套房平面

0   1     3 m

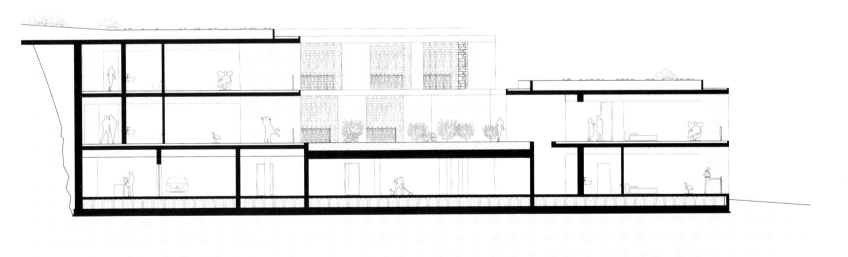

cross section
横截面

0    2     5 m

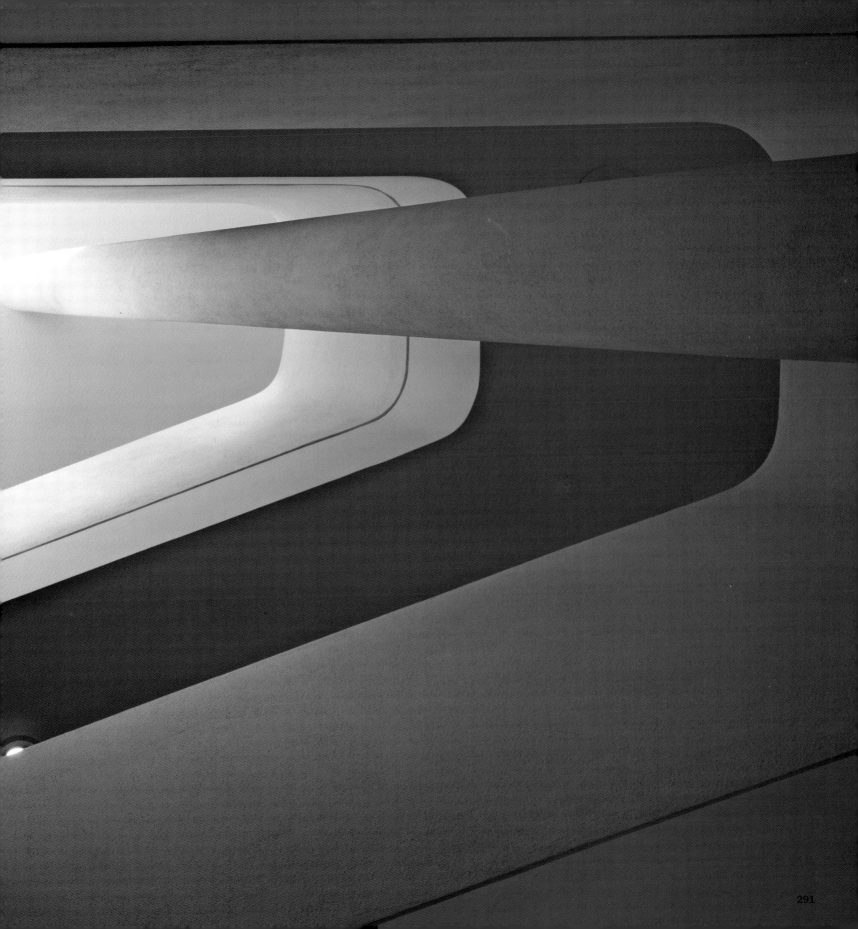

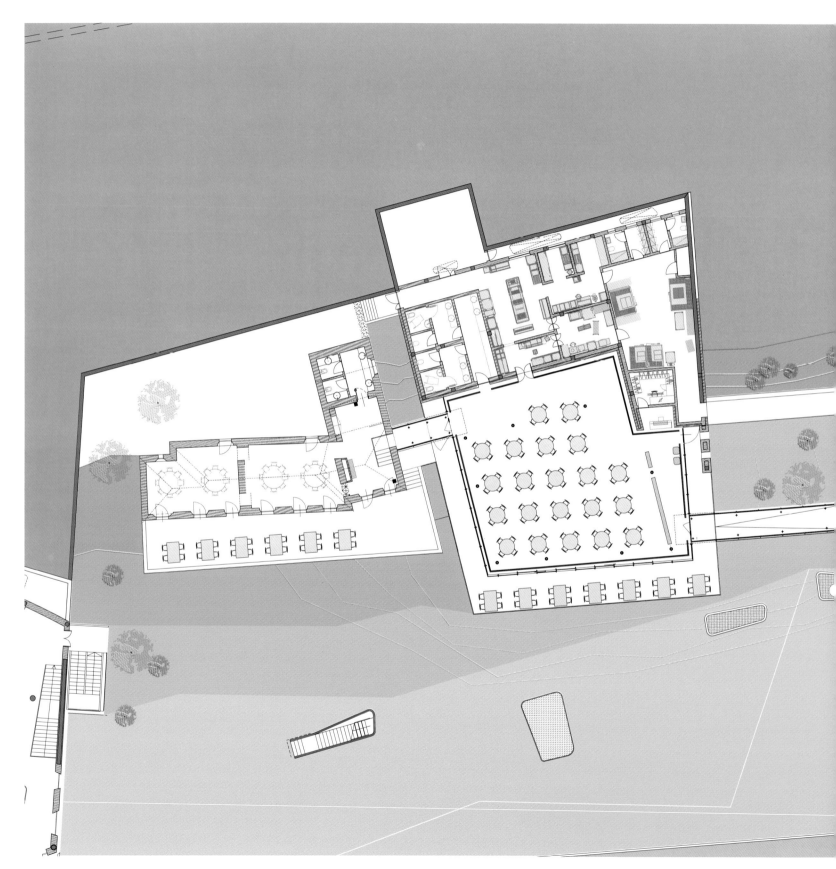

plan level + 10.60 m
水平高度+10.60米平面

0   2   5 m

# PADUA MASTERPLAN

## Padua, Italy, 2008-2009

地点　帕多瓦，意大利
项目　总体规划
客户　I.F.I.P.
规划　2008年-2009年
占地面积　61,405平方米
建筑面积　48,506平方米
体量　156,400立方米
承包商　I.F.I.P.

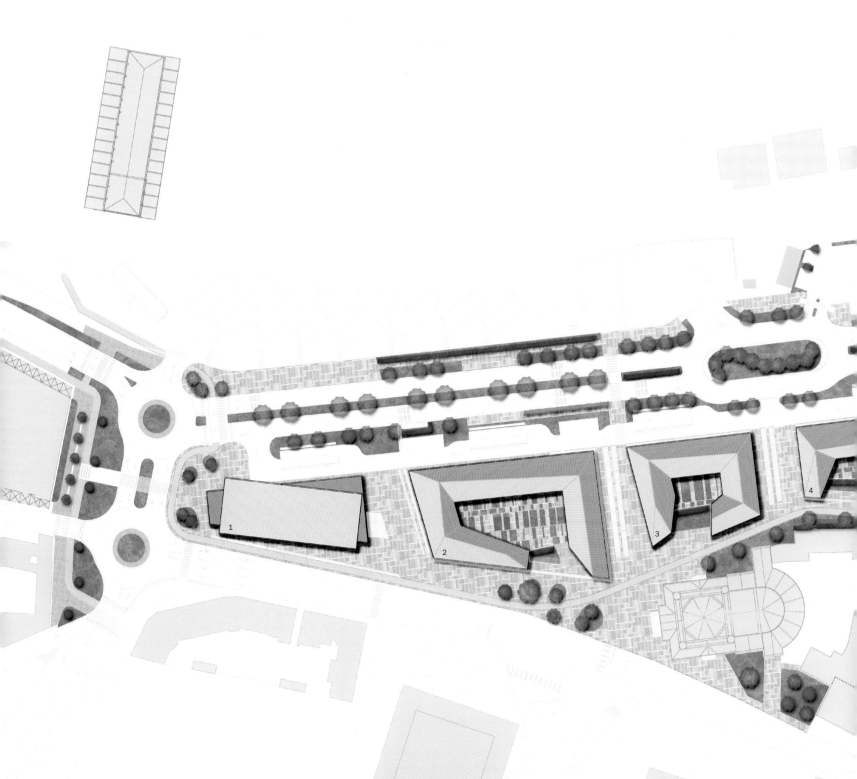

1. H building, not residential
   H号建筑，非居住

2. R1 building, not residential
   R1号建筑，非居住

3. R2 building, residential
   R2号建筑，居住

4. C building, residential
   C号建筑，居住

5. A1 building, not residential
   A1号建筑，非居住

6. A2 building, not residential
   A2号建筑，非居住

7. B building, not residential
   B号建筑，非居住

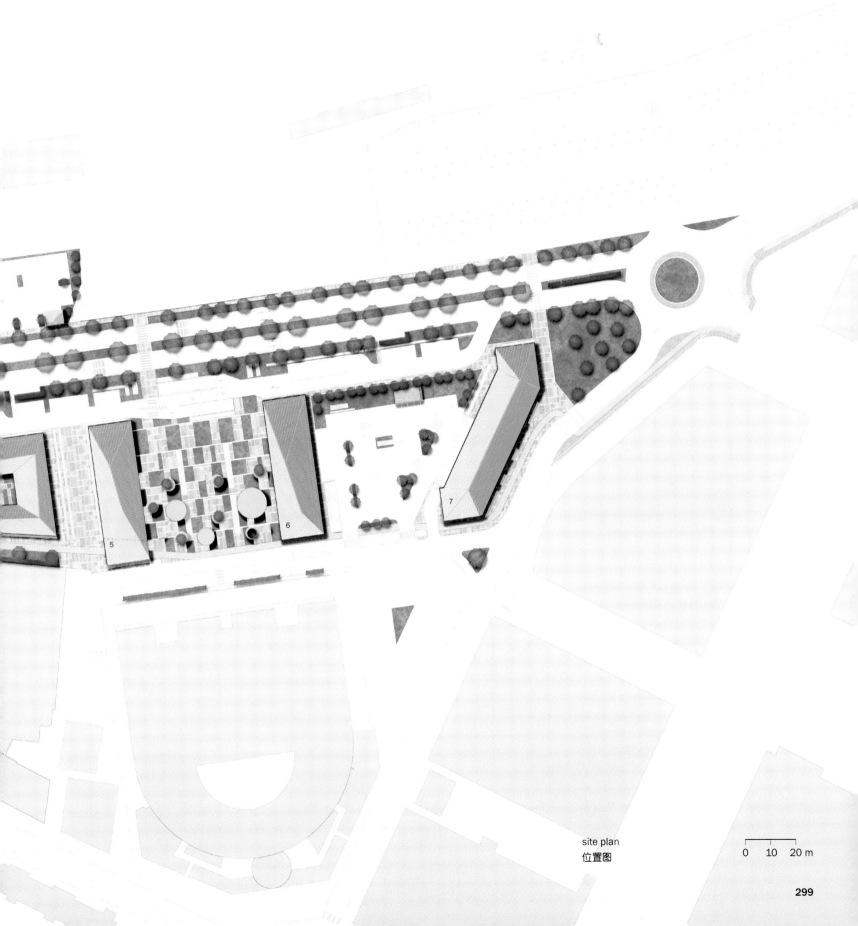

site plan
位置图

0　10　20 m

urban preliminary study
城市初步规划

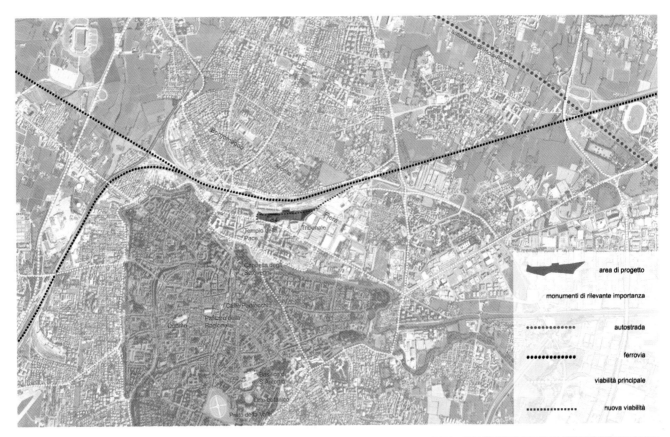

area di progetto

monumenti di rilevante importanza

............. autostrada

•••••••••••••• ferrovia

viabilità principale

▪▪▪▪▪▪▪▪▪▪ nuova viabilità

relevant buildings
相关建筑

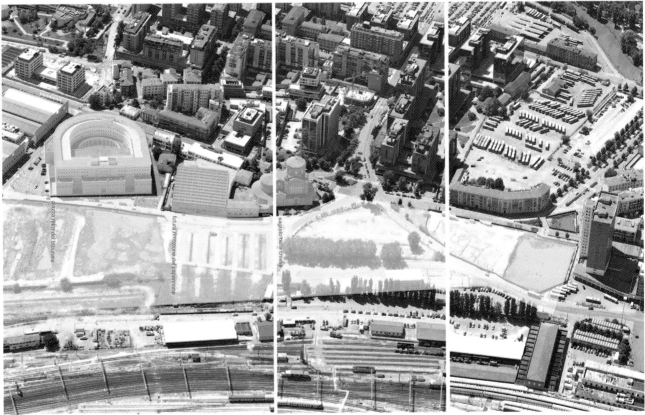

R2 building, type plan
R2号建筑，平面类型

0　　　5　　　10 m

R1 building, type plan
R1号建筑，平面类型

0  5  10 m

305

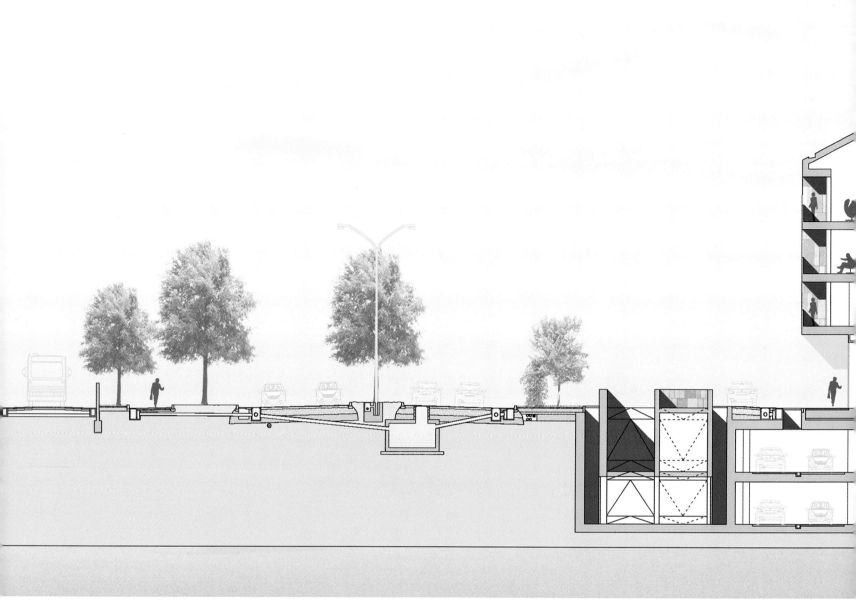

cross section
横剖面

0   2   5 m

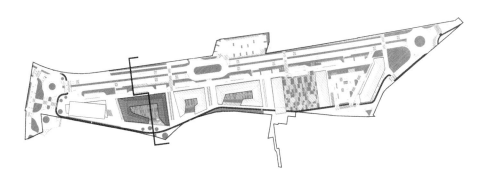

# MERAVIGLIOSA ISLAND

**marina residence**
**The world, Dubai, United Arab Emirates**
**2009 - in progress**

**地点** 世界岛，杜拜，阿拉伯联合酋长国
**项目** 住宅，酒店
**客户** Rajaa Trading & Investiments
**规划** 2009年
**建设** 在建
**占地面积** 40,642平方米
**建筑面积** 16,782平方米

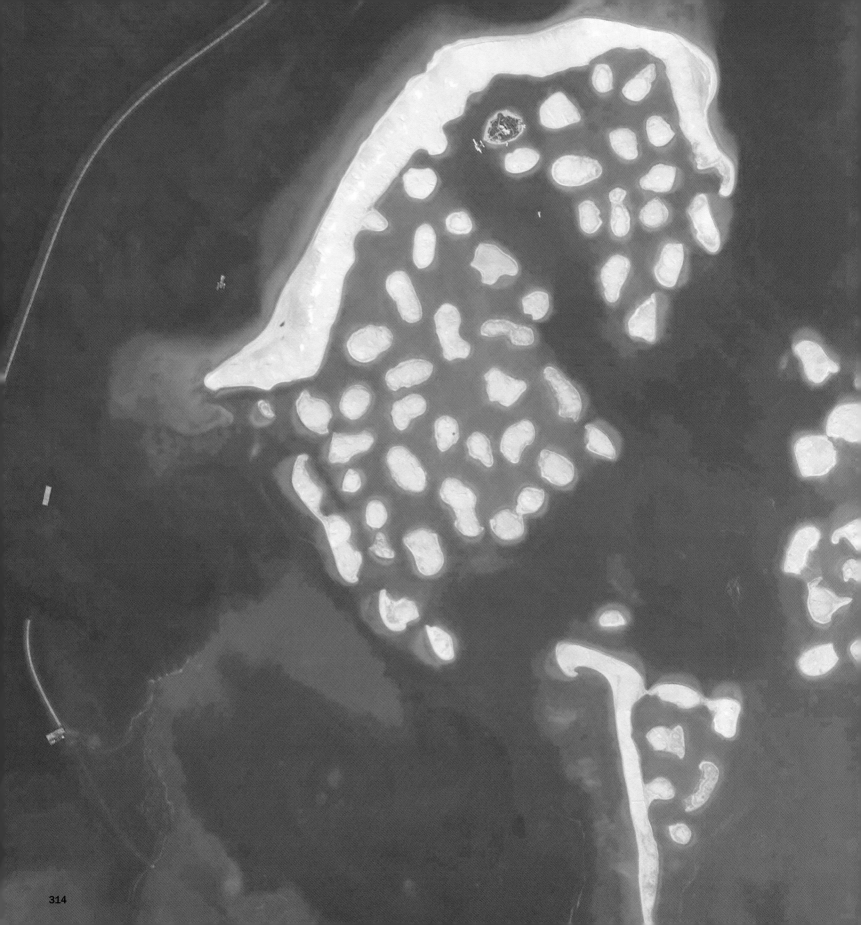

314

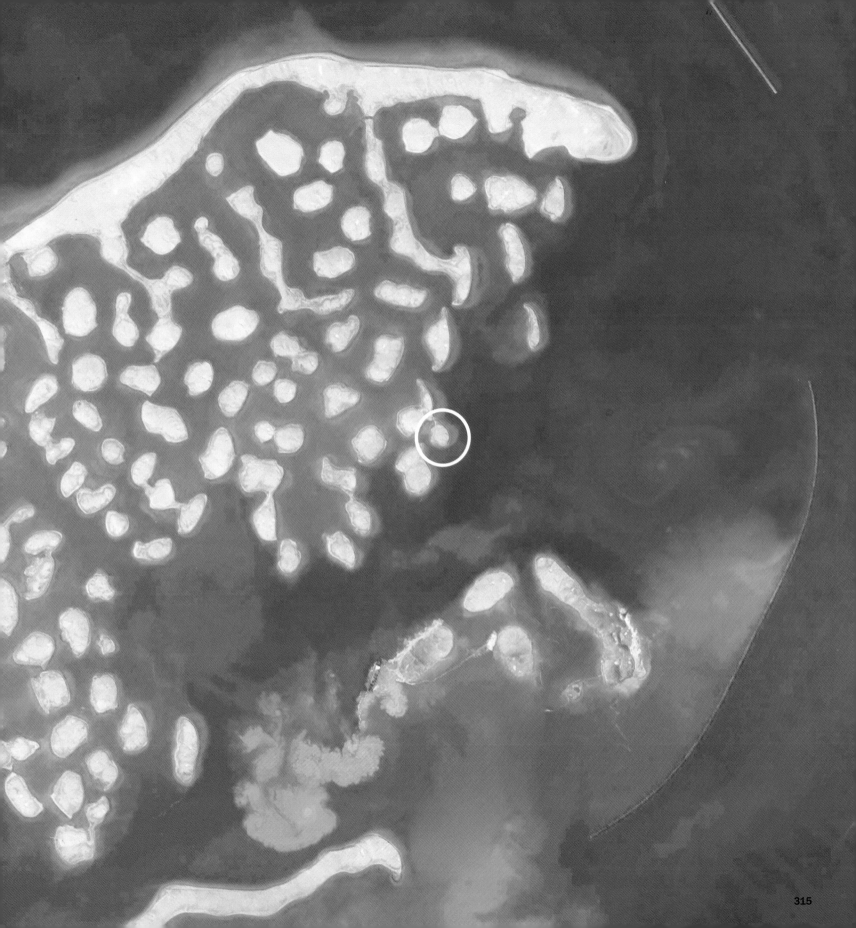

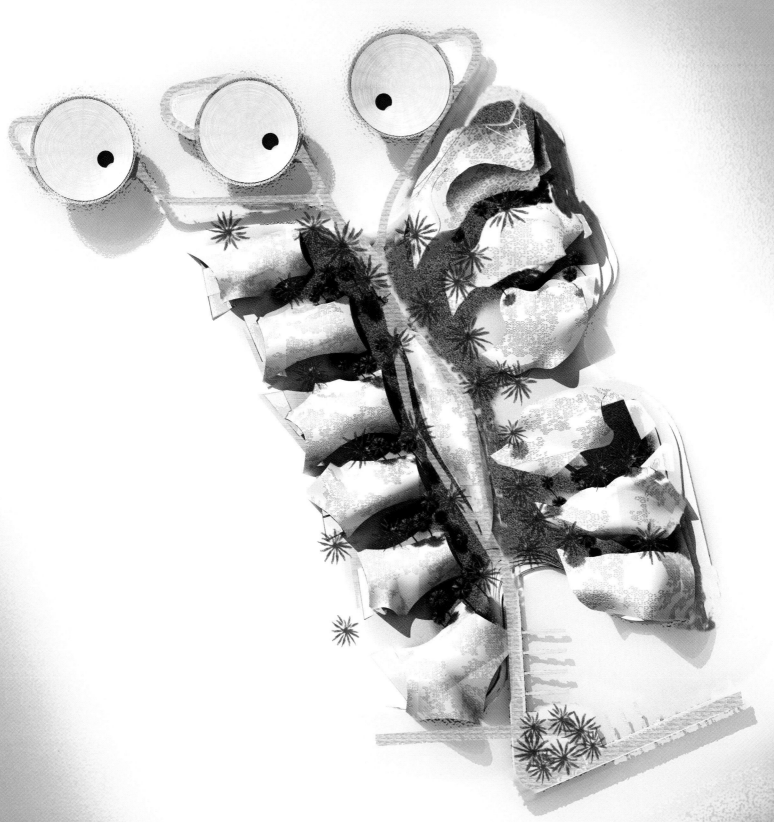

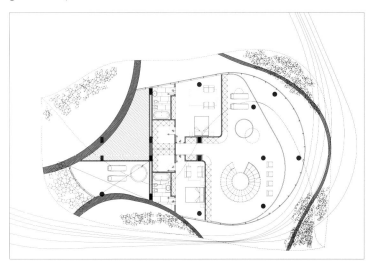

ground floor plan　一层平面图

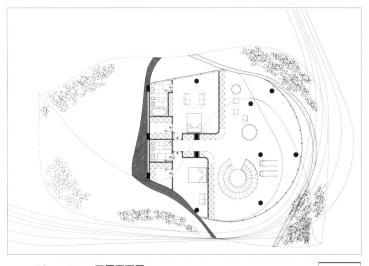

first floor plan　二层平面图

second floor plan　三层平面图

0　　5 m

## OCEAN VILLA
## 海岸别墅

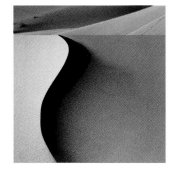

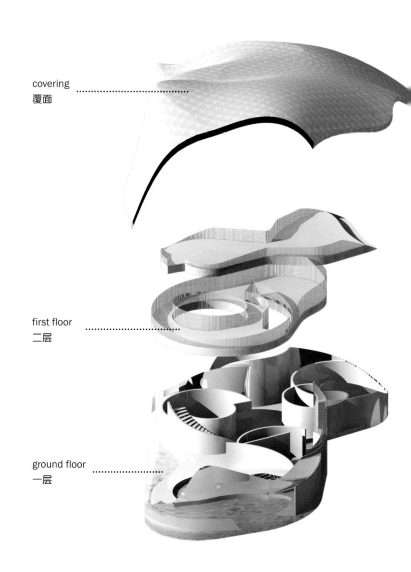

covering
覆面

first floor
二层

ground floor
一层

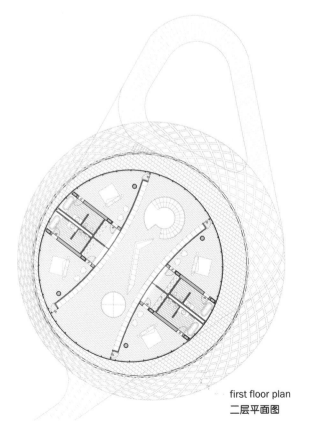

ground floor plan
一层平面图

first floor plan
二层平面图

0    5 m

# WATER VILLA
# 海岸别墅

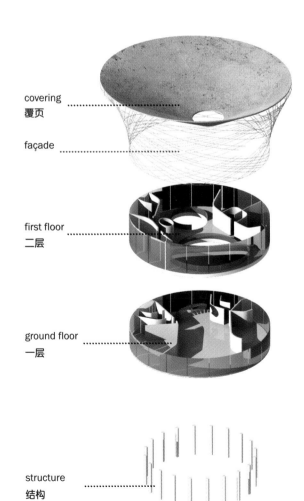

covering
覆页

façade

first floor
二层

ground floor
一层

structure
结构

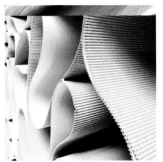

# BEACH VILLA
# 海滩别墅

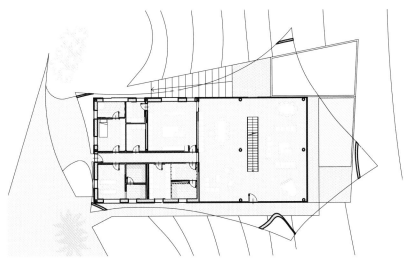

ground floor plan　一层平面

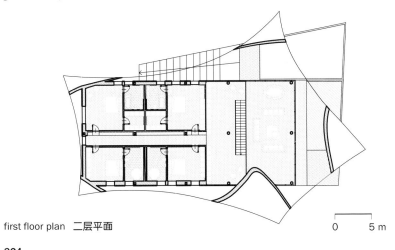

first floor plan　二层平面

0　　5 m

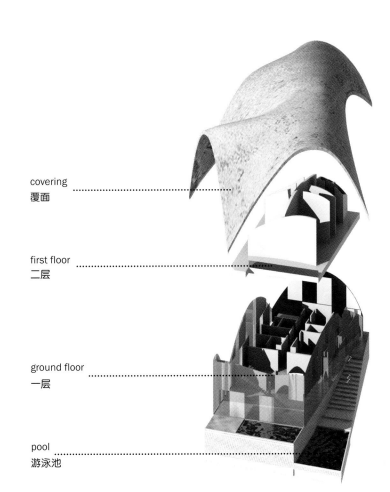

covering
覆面

first floor
二层

ground floor
一层

pool
游泳池

# LILING WORLD CERAMIC ART CITY

## Liling, China, 2010 - under construction

地点　醴陵，中国
客户　建设单位
项目　醴陵世界陶瓷艺术城
规划　2010年
建设　建设中
占地面积　110000平方米

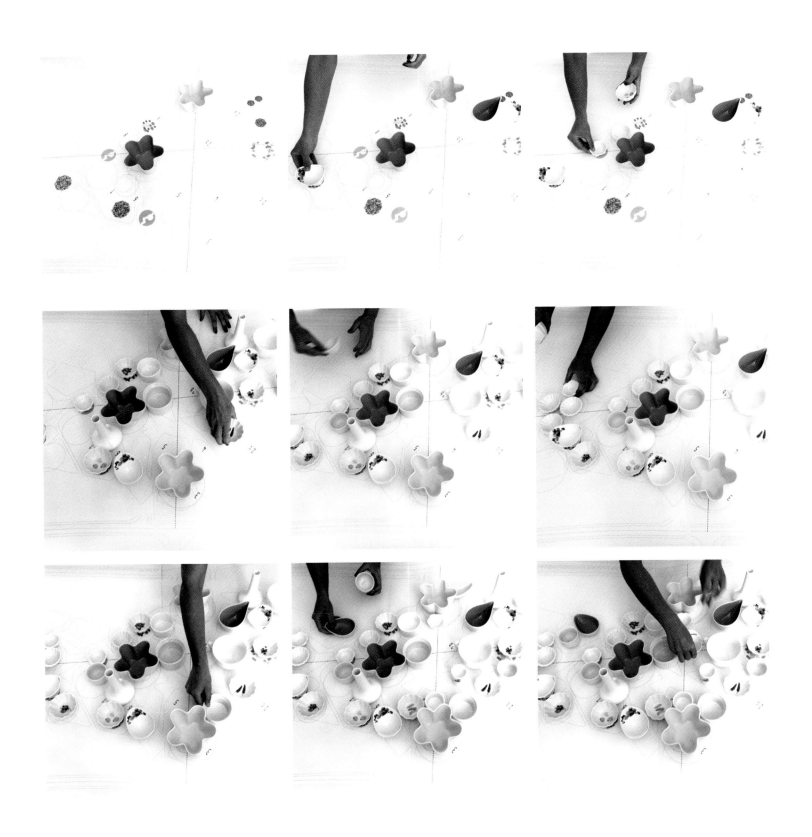

site plan / 位置图

0    25    50 m

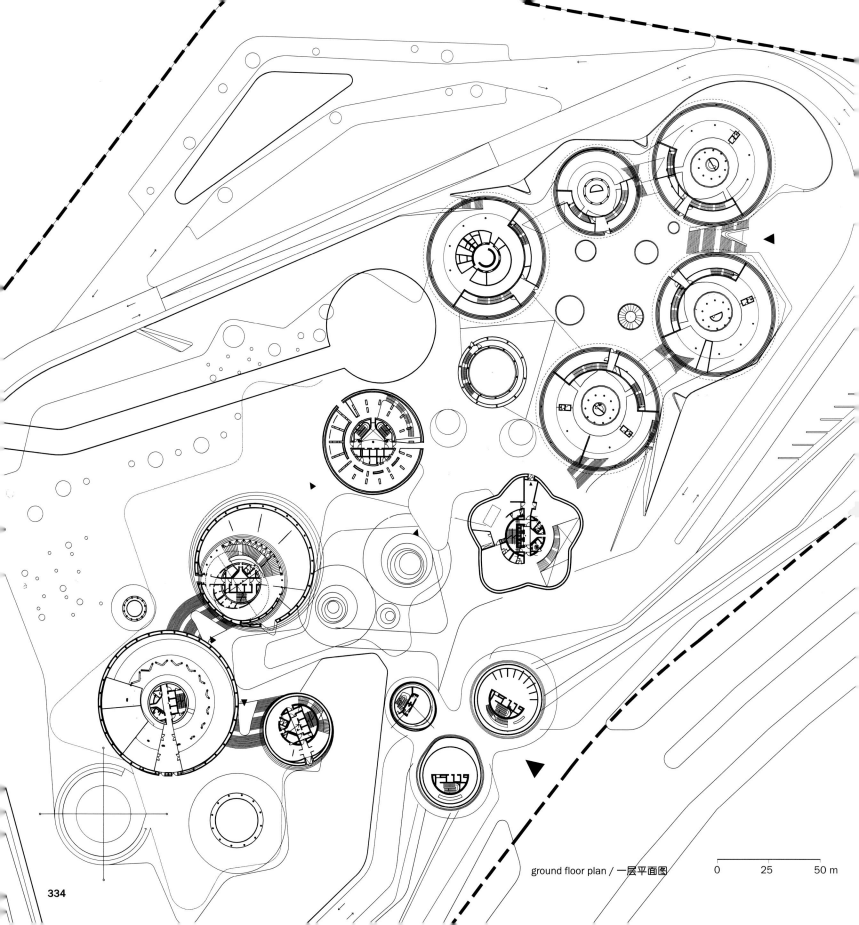

ground floor plan / 一层平面图

0      25      50 m

# LVBO CORE CLUSTER AREA

## Zhengzhou, China, 2011

地点　郑州，中国
客户　郑州中牟产业园区管委会
规划　2011年
占地面积　22445163 平方米
建筑面积　16564269 平方米
承包商　郑州中牟产业园区管委会

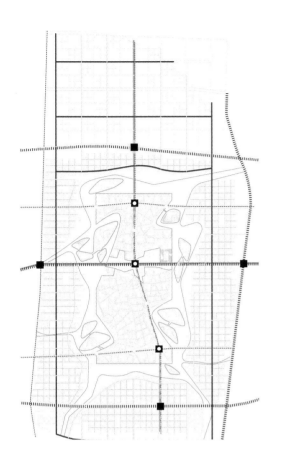

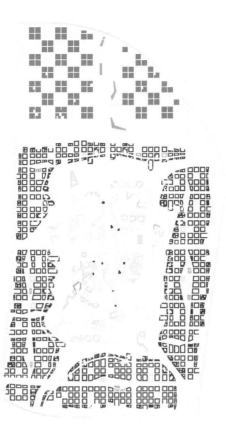

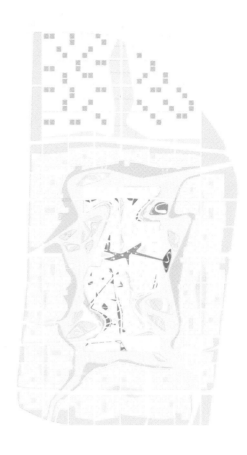

GENERAL PLAN: TRAFFIC SYSTEM PLANNING AND
STATION SYSTEM
总体规划：规划和交通系统站系统

—— intercity railway / 城际铁路

—— metro line / 地铁线

· · · national highway / 国家高速公路

—— people mover / 公交系统

—— secondary truck road / 次干道

|||||  first - level main road / 一级主干道

· · · · · Second - level main road / 二级主干道

—— pedestrian path / 人行道

—— internal road (city) / 内部道路（城市）

—— branch way / 支路

○ intercity railway station / 城际铁路站

people mover station / 公交车站

■ metro station / 地铁站

GENERAL PLAN: LAND USE AND FUNCTIONS
总体规划：土地使用及功能

■ residential area / 住宅区

medical treament and health land / 医疗卫生用地

■ living service land / 生活服务用地

first-level industry land / 一类工业用地

education and scientific research land / 教育科研用地

culture and entertainment land / 文化娱乐用地

commerce and finance land / 商业金融业用地

administrative office land / 行政办公用地

logistics service land / 物流服务用地

■ sports land / 体育用地

GENERAL PLAN: GREEN SYSTEM
总体规划：绿化系统

urban green / 城市绿地

ecological conservation area / 生态保护区

bridge park / 桥体公园

forest park / 森林公园

green barrier / 隔离绿地

industrial green roof / 工业屋顶绿化

lake park / 滨水公园

linear public park / 线状公园

public green fitted / 公共绿色铺装

residential green roof / 住宅屋顶绿化

sustainable park / 可持续公园

0    50    100 m

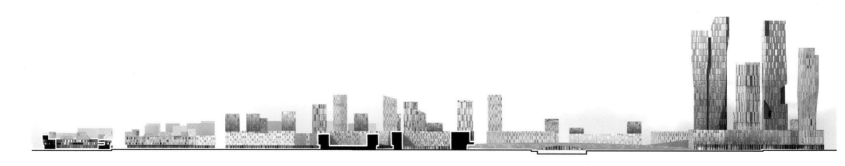

cross section
横剖面

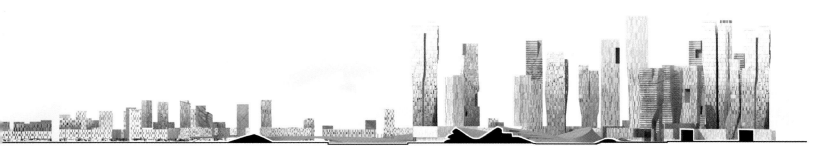

0　　　　　50　　　　　100 m

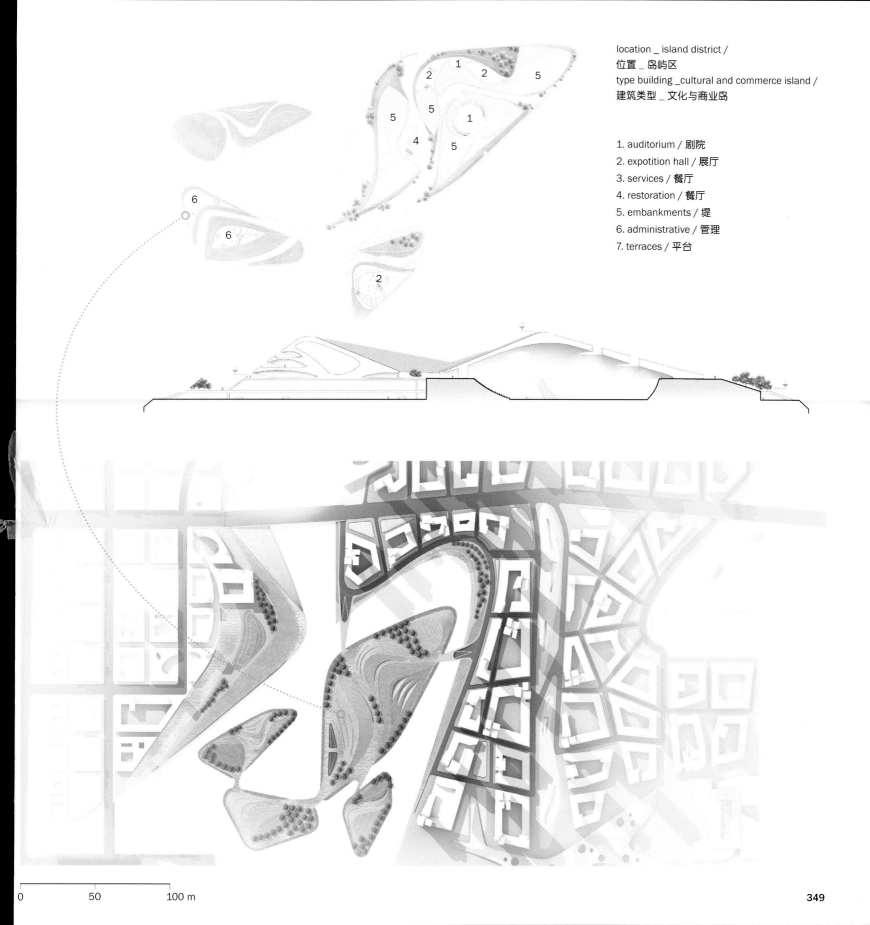

location _ island district /
位置 _ 岛屿区
type building _ cultural and commerce island /
建筑类型 _ 文化与商业岛

1. auditorium / 剧院
2. expotition hall / 展厅
3. services / 餐厅
4. restoration / 餐厅
5. embankments / 堤
6. administrative / 管理
7. terraces / 平台

0    50    100 m

图书在版编目（CIP）数据

可持续性地标建筑 ： 汉英对照 ／ 石大伟 主编. --北京 ： 中国林业出版社，2012.6

ISBN 978-7-5038-6603-6

Ⅰ. ①可… Ⅱ. ①石… Ⅲ. ①建筑设计－作品集－世
界－现代 Ⅳ. ①TU206

中国版本图书馆CIP数据核字（2012）第094637号

可持续性地标建筑（上）                                                石大伟  主编
-----------------------------------------------------------------------

责任编辑：李  顺
出版咨询： （010）83223051
-----------------------------------------------------------------------
出 版：中国林业出版社（100009 北京西城区德内大街刘海胡同7号）
印 刷：北京时捷印刷有限公司
发 行：新华书店北京发行所
电 话：（010）83224477
版 次：2012年6月第1版
印 次：2012年6月第1次
开 本：787mm×1092mm 1／12
印 张：76
字 数：200千字
定 价：880.00元（上、中、下册）